essentials

Springer essentials

Springer essentials provide up-to-date knowledge in a concentrated form. They aim to deliver the essence of what counts as "state-of-the-art" in the current academic discussion or in practice. With their quick, uncomplicated and comprehensible information, *essentials* provide:

- an introduction to a current issue within your field of expertise
- an introduction to a new topic of interest
- an insight, in order to be able to join in the discussion on a particular topic

Available in electronic and printed format, the books present expert knowledge from Springer specialist authors in a compact form. They are particularly suitable for use as eBooks on tablet PCs, eBook readers and smartphones. *Springer essentials* form modules of knowledge from the areas economics, social sciences and humanities, technology and natural sciences, as well as from medicine, psychology and health professions, written by renowned Springer-authors across many disciplines.

Wolfgang Kemmler · Michael Fröhlich · Christoph Eifler

Whole-Body Electromyostimulation

Effects, Limitations, Perspectives of an Innovative Training Method

Wolfgang Kemmler
Institute of Medical Physics,
Friedrich-Alexander-University
Erlangen-Nürnberg and Institute
of Radiology, University-Hospital
Erlangen
Erlangen, Germany

Michael Fröhlich
Department of Sports Science
Rheinland-Pfälzische Technische
Universität Kaiserslautern-Landau
Kaiserslautern, Germany

Christoph Eifler
German University for Prevention
and Health Management
Saarbrücken, Germany

ISSN 2197-6708 ISSN 2197-6716 (electronic)
essentials
ISSN 2731-3107 ISSN 2731-3115 (electronic)
Springer essentials
ISBN 978-3-031-56710-0 (eBook)
https://doi.org/10.1007/978-3-031-56710-0

This Springer imprint is published by the registered company Springer Nature Switzerland AG
The registered company address is: Gewerbestrasse 11, 6330 Cham, Switzerland

Paper in this product is recyclable.

What you can find in this *essential*

- Introduction to whole-body electromyostimulation (WB-EMS)
- Methods, procedures and parameters of WB-EMS
- Recommendations for WB-EMS application, contraindications and safety aspects
- Evidence on effects of WB-EMS application on functional and health-related outcomes
- WB-EMS from the commercial point of view

Preface

Whole-body electromyostimulation (WB-EMS) is a training technology that is gaining increasing attention in several areas of application. Particularly the unique feature that most if not all the main muscle groups are simultaneously stimulated, albeit with dedicated intensity for each and without the need to apply additional load, might explain the attractiveness of WB-EMS for many people. In essence, the correspondent time efficiency and joint friendliness and adding the aspects of non-voluntary stimulation and high customization of WB-EMS application indicate its broad applicability in the domain of performance, fitness/function, body composition and health-related outcomes. While various cohorts and outcomes have been addressed by WB-EMS research, many gaps in research are still present. Apart from effectiveness, safety aspects are of particular importance considering the comprehensive, in excess supra-maximum, character of WB-EMS. Therefore, the present essential includes the updated recommendations on safe and effective WB-EMS and the revised contraindications on WB-EMS. Lastly, the commercial aspect of and the outlook for WB-EMS will be addressed. We focus on the German market with its nearly 20 years of experience of WB-EMS application that might serve as a blueprint for further trends and development in other countries. In summary, this essential aims to provide a quick overview of different aspects of WB-EMS application and thus offer a first impression on features, benefits and limitations of this innovative training technology.

February 2024

Wolfgang Kemmler
Michael Fröhlich
Christoph Eifler

Contents

1 Introduction

Whole-body electromyostimulation (WB-EMS) is a training technology of German origin that is enjoying increasing popularity in various cohorts, settings and fields of application. Apart from its joint-friendly feature, the main reason for its attractiveness is definitely the high time effectiveness of standard WB-EMS programs. In commercial WB-EMS settings, there is usually only one 20-min session per week — with predominantly three sessions in two weeks the training frequency is widely similar in most scientific studies. The vast majority of providers (and researchers) apply WB-EMS as a resistance type exercise training. Apart from the low training frequency, this includes moderate to high (impulse) intensity, intermitted application (4–6 s of stimulation—4 s rest) and low duration of the session (20 min). Although WB-EMS could also applied in a specification-related endurance exercise (i.e. higher volume/lower intensity), due to the loss of time efficiency only few commercial providers rely on these concepts. Indeed, the unique selling point of high time efficiency makes WB-EMS interesting for a large variety of users. This particularly refers to people with limited time resources, athletes of non RT disciplines, but also older people who additionally benefit from the WB-EMS aspect of joint-friendliness, close interaction and supervision of the present concepts.

The motivation to use WB-EMS covers all facets and goals of performance, fitness and health-related outcomes. According to an customer survey from 2017 (EMS-Training.de 2017) the rationale for commercial non-medical WB-EMS application is predominately physical attractiveness, physical fitness and reduction of low-back pain. Although no updated data are available, we feel that there has been a pronounced shift from physical attractiveness to functional and health-related aims during the last few years. Correspondingly, the number of clients of

W. Kemmler et al., *Whole-Body Electromyostimulation,* essentials,
https://doi.org/10.1007/978-3-031-56710-0_1

an advanced age have increased; and this requires a more dedicated addressing of this cohort by commercial providers.

In this essential, we aimed to characterize WB-EMS technology in the first two chapters. Chapter 3 focuses on safety aspects of WB-EMS and provides recommendations for safe and effective application. Chapter 4 presents a brief review of cohorts and outcomes addressed by WB-EMS studies so far and evidence for WB-EMS effects. Lastly, Chap. 5 takes a look at the situation, trends and developments in the WB-EMS market in order to introduce the reader to the commercial application perspective.

1.1 Definition of Whole-Body Electromyostimulation

Whole-body electromyostimulation (WB-EMS) is defined as "Simultaneous application of electric stimuli via at least 6 current channels or participation of all major muscle groups, with a current impulse effective to trigger muscular adaptations" (Fig. 1.1).

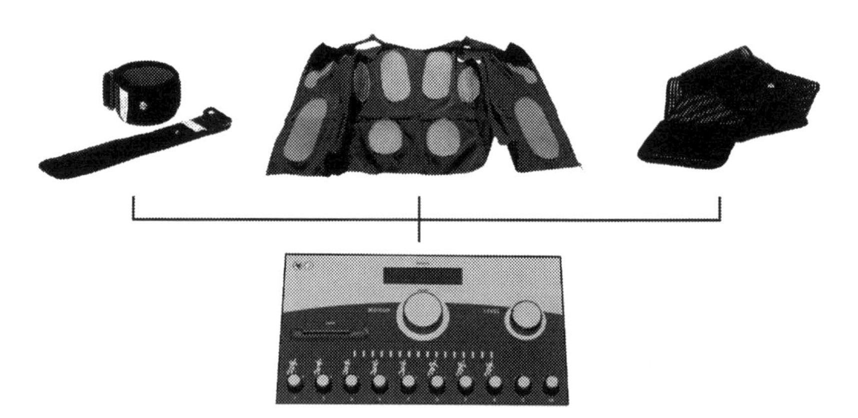

Fig. 1.1 WB-EMS device with equipment (electrode vest, gluteal, thigh and upper arm belts)

1.2 Classification and Categorization of WB-EMS

WB-EMS is a technology based on neuromuscular electrical stimulation (NMES). Neuromuscular electrical muscle stimulation (NMES) and electromyostimulation (EMS) trigger contractions of muscle groups or individual muscles with the help of an electrical stimulus independent of innervated and denervated muscle status. Muscles are not stimulated directly but by the innervated nerves. Of importance, there is increasing evidence that cortical, subcortical and segmental neural networks are activated in a way that is similar to active training sequences with voluntary activation (review in (van Kerkhof 2022)). Thus, the frequently applied limitation that NMES/EMS provides simple effects on muscle parameters without any neural (functional) aspects have to be revised.

1.3 Differences to Local EMS

Differences between WB- and local (N)EMS application are quite obvious. While widely known local EMS focuses on muscle stimulation by single electrodes placed on dedicated trigger points of the muscle, WB-EMS covers all the main muscle groups simultaneously using vests with large stimulation areas and belt electrodes for the extremities. Altogether up to 2800 cm^2 of area can be addressed when stimulating both calves, thighs, gluteals, lower and upper back, latissimus dorsi, chest, abdominals and both upper arms (Fig 1.1). Correspondingly, the use of local EMS focuses predominately on the therapy of regional lesions and injuries while WB-EMS covers a much larger area of application.

WB-EMS is defined as the "simultaneous application of electric stimuli via at least six current channels or participation of all major muscle groups, with a current impulse effective to trigger muscular adaptations" (Kemmler et al. 2020).

Apart from the greater stimulation area during WB-EMS, several other differences are evident, not necessarily related to WB-EMS in general, but more to the specification applied for most commercial WB-EMS devices.

- In contrast to local electromyostimulation with its possibility to address any muscle by adhesive electrodes, WB-EMS electrodes are firmly embedded in the vest and belts. Most systems provide 1–2 free current channels for optionally addressing additional muscle groups by large sized electrodes.

- During WB-EMS of the extremities, agonist and antagonist muscles as well as neighboring muscle groups are stimulated simultaneously via the ring-shaped arm and leg cuffs. It is not possible to specifically regulate impulse parameters for individual muscle groups below the cuff electrode.
- In addition, due to the bipolar impulse setting of most WB-EMS devices, most electrodes (i.e. legs, arm, chest, upper back, latissimus) are attached to the opposite sides, i.e. the left and right side of the muscle group are addressed with the same stimulation specifications. This prevents the specific stimulation of the weaker side with other impulse specifications, for example. However, for the extremities at least, this limitation can be overcome by placing both cuff electrodes on the same extremity and using free electrodes for the opposite extremity in order to regulate intensity for the individual extremities separately.
- The electrodes applied during WB-EMS are much larger compared with the small adhesive electrodes used by local EMS. This enables a more comprehensive stimulation of large muscle groups. The large electrode surfaces integrated into the WB-EMS vests and cuffs further prevent the time-consuming and complicated placement of local electrodes on dedicated trigger points.
- While most local EMS settings focus on static application, the majority of WB-EMS protocols use dynamic protocols—albeit predominately (Sect. 2.3) with low voluntary exercise intensity.

1.4 Commercial Non-Medical Versus Medical WB-EMS

The last few years have seen a new development in commercial setting that reflects the increasing interest of vulnerable cohorts in WB-EMS as a joint-friendly, closely supervised and safe alternative to conventional exercise. Although not yet fully established in the healthcare system, medical WB-EMS focuses particularly on health-related outcomes in older and/or functionally impaired people. Recently, a definition of medical WB-EMS was provided (Berger et al. 2022) that includes the specifications of (1) primarily therapeutic intervention (2) based on an existing diagnosis (3) provided by qualified medical-therapeutic personnel (4) using medical devices in compliance with current guidelines (5).

Methods and Procedures of WB-EMS 2

2.1 Basics of Neuromuscular Electrostimulation via WB-EMS

As already noted, the term "muscle stimulation" is a misleading one in that the threshold for exciting motor nerves is significantly lower than that of the muscle in question. In WB-(N)EMS, the superficially applied impulse first stimulates the nerve which then causes muscle contraction via the alpha motoneurons and the motor end plates. Leaving to one side the multitude of possible applications of electrical currents and concentrating on the currently established technology: WB-EMS generally uses bipolar (biphasic) stimulation currents in the low frequency (0–1000 Hz) and (modular) medium frequency range (>1000 Hz to <1 MHz). Most protocols apply amplitude modulation, i.e. intermitted impulse protocols with short breaks between the impulse sequences. The most important impulse parameters of WB-EMS are outlined below.

> Similar to conventional exercise protocols, the close interaction between the large number of training and stimulation parameters has to be carefully considered when designing a WB-EMS protocol to successfully address a given outcome. This requires sufficient expertise in the relevance and effects of the parameters described below.

W. Kemmler et al., *Whole-Body Electromyostimulation,* essentials,
https://doi.org/10.1007/978-3-031-56710-0_2

2.2 Impulse Parameters

2.2.1 Impulse Type

The vast majority of commercial WB-EMS devices apply biphasic (bipolar) impulse. In contrast to a monophasic (monopolar) pulse, there is a constant change in the direction of impulse flow ("alternating current"). To stimulate large muscle areas, electrodes of the same size are attached to the corresponding muscles on the right and left sides of the body. Thus, in contrast to local application, stimulation is not predominantly punctual, but extensive and in the sense of a transverse flow.

2.2.2 Impulse Frequency

The impulse frequency describes the number of individual pulses per second (in "Hertz", Hz). There is an ongoing disagreement between advocates of low-frequency[1] WB-EMS applications in the 7–120 Hz range and supporters of a modulated medium impulse frequency range as to the "most favorable impulse frequency". Modulated medium-frequency currents, i.e. medium-frequency based currents ($\approx$2 kHz) with a low-frequency modulation (0.5–250 Hz), might offer the advantage of a lower sensitive load (van Kerkhof 2022) and are possibly associated with a greater "depth of stimulation". A disadvantage of this method is the pronounced "high-frequency fatigue", which to a certain extent contradicts the aims of muscle training (Stefanovska and Vodovnik 1985). However, the evidence for the fundamental superiority of a method based on strength development, for example, as the predominately addressed study outcome is rather limited. First of all, both methods augment the maximum forces generated by voluntary contraction (Pinfildi et al. 2018). Stefanovska and Vodovnik (1985), who compared both methods with muscle strength of the leg extensors as the primary outcome, reported (non-significantly) higher strength gains (25% vs. 13%) after low frequency (25 Hz) compared with medium-frequency alternating current. In parallel, Bircan et al. (2002) reported similar significant effects (vs. non-training control) after 3 weeks of low (80 Hz) vs. modulated medium-frequency (80/2500 Hz) on quadricep strength. The authors further reported no differences in perceived discomfort. Data on acute strength development show comparable (Aldayel et al. 2010), superior (Laufer and Elboim 2008) or inferior effects (Pinfildi et al. 2018)

[1] 0–1000 Hz (Edel 1991).

of low- vs. medium-frequency electrical stimulation. To our best knowledge, there is only one clinical study that focuses on medium-frequency electrical stimulation (Lukashevich 2020); and so practically all the data and study results in the present essential refer to low frequency stimulation. Looking at the low frequency domain, there is a close correlation between impulse frequency and contraction force. The range of ≥50–90 Hz is considered to be particularly favorable for muscle and strength development, while lower frequencies are used for aims related to regeneration, blood circulation and endurance training. However, the effects of impulse frequency on dedicated outcomes have to be considered in context with the adjuvant training and impulse parameters. Further, comparative studies have not always confirmed superiority for frequencies above 50 Hz for strength development. Correspondingly, a 10-week study with a standard WB-EMS protocol (Sect. 2.3) showed no significant differences at stimulation frequencies of 25 Hz vs. 85 Hz on strength and performance parameters in untrained subjects (Berger et al. 2020).

2.2.3 Impulse Width

The impulse width describes the duration or exposure time of a single pulse in μs. It basically depends on the time required to excite the motor nerve fiber (chronaxy), which varies considerably. The smaller the pulse width, the higher the pulse amplitude ("intensity") must be to generate an over-threshold stimulus. With a moderately high pulse width (300–400 μs), the stimulus generally penetrates deeper into the tissue and recruits more motor units (van Kerkhof 2022), while low pulse widths remain superficial and high (≥500 μs) pulse widths are perceived as painful.

2.2.4 Impulse Rise

The impulse rise characterizes the duration until the impulse reaches its maximum. Immediate, rectangular impulse increases and decreases usually show the most favorable effects on muscle strength and performance parameters and are primarily used in WB-EMS for sufficiently trained individuals. With high impulse intensity, a rectangular pulse can be painful, even with active voluntary pre-tensioning. Depending on the impulse duration and training aim, a ramp-shaped rise and decrease with a "creeping" effect in the 0.4–0.6 s range is thus recommended for novices and vulnerable applicants.

2.2.5 Impulse Duration

The impulse duration is characterized by the length of the impulse stimulus. In principle, a distinction is made between continuous protocols with WB-EMS application over the entire session and intermitted (on–off) protocols. In line with basic sports science, long or even prolonged impulse durations correlate negatively with the impulse intensity and are considered more effective for (strength) endurance. The strength-oriented standard EMS protocols from commercial providers predominantly use short impulse durations (4–6 s) interspersed with impulse pauses (4 s). The so-called "duty cycle" ("on–off" ratio) describes the stimulus density. A duty cycle of 50% would therefore be achieved by an impulse duration of 4 s interrupted by an impulse break of 4 s.

2.2.6 Impulse Intensity (or Amplitude)

Although there is a close interaction between the strain and impulse parameters, impulse intensity might be particularly important for the effectiveness of WB-EMS on a given outcome. Similar to resistance exercise, a distinction should be made between absolute and relative (exercise) intensity. Absolute impulse intensity in milliamperes (mA) is not suitable for specifying intensity in WB-EMS, however. A more pragmatic and more individual approach is intensity specification via the rate of maximum voluntary contraction (%MVC) and in particular the rate of maximum tolerable stress (%1MT). To realize relevant muscular adaptations, an impulse intensity of >50% MVC is recommended. Although some approaches (e.g. Kim and Jee 2020b; Kim et al. 2021; Park, Min et al. 2021; Park, Park et al. 2021) determined the maximum tolerable impulse intensity (1MT) for each of the muscle groups, the intensity specification at least in non-athletic settings is not based on this — when properly applied — rather painful procedure. Considering further that maximum tolerable stimulus intensity has been reported to be closely related to rhabdomyolysis at least in novice applicants, the common practice of specifying impulse intensity via the rate of perceived exertion tends to be applied. Most popular, the Borg CR 10 scale (Borg and Borg 2010) has become established as the standard specification, with "5" already being considered as "hard" and "10" as extremely hard. A frequently stated limitation of the RPE approach is the need for body awareness, a calibrated sense of exertion and a high degree of compliance (at least at higher impulse intensity levels) of the applicant. In particular in this context, the guidance and supervision of an experienced trainer seems to be crucial for ensuring an appropriate impulse intensity for each body region/muscle group. Of note, due to the adaptation that sets in during the training session, the impulse intensity (mA) per muscle group has to be increased successively in order to keep the absolute intensity constant (Sect. 3.2.4). This indicates that a consistent progression of impulse intensity is inherent in this method.

2.2.7 Length of WB-EMS Application

In principle, the duration of a WB-EMS training session depends primarily on the composition of the impulse parameters and therefore on the training aim. As stated, most commercial providers and researchers apply a resistance-type approach with high impulse intensity (RPE 6–8), immediate impulse boosts, and short bouts of stimulation (4–6 s) intermitted by 4 s of impulse break (Beier et al. 2024). In summary, this protocol allows a short total duration of just 20 min. Nevertheless, when addressing aerobic or endurance-related training aims a longer duration of the WB-EMS session with lower stimulus intensity and continuous stimulation might be preferable. However, since a unique selling point of WB-EMS is its time efficiency, longer WB-EMS sessions might reduce the attractiveness for applicants with limited time resources.

2.2.8 Kind of Application

Regardless of the stimulation parameters, WB-EMS can be performed in a passive or active style, i.e. with or without voluntary movements/exercises. Although not undisputed (Stephan et al. 2023), study data (Kemmler et al. 2015) indicate more favorable effects on maximum strength and body composition of active WB-EMS performed with (vs. "without") low intensity movements — at least in older people. From a functional point of view, active WB-EMS training appears to be more effective anyway, as the movements and active muscle control can improve coordination. Reviewing the literature on "active" WB-EMS approaches (Beier et al. 2024), the level of voluntary activation differs considerably. While about half of the studies applied movements/exercises with "subthreshold" intensity (Bloeckl et al. 2022; Kemmler et al. 2016; Kemmler et al. 2017; Micke et al. 2021) others (e.g. Akçay et al. 2022; Berger et al. 2020; Park, Min et al. 2021) combined WB-EMS with resistance exercises. The concept of "superimposed" WB-EMS, which is predominately applicable for athletic or competitive sports, relies on sport-specific or discipline-relevant voluntary high intensity muscle activation, superimposed by additional WB-EMS application. In diametral contrast, the health and fitness sector usually works with low-intensity voluntary muscle activation and moderate to high intensity impulse application — i.e. "the electric current does the work".

2.2.9 Supervision

Although physical supervision by a certified trainer should be considered a "must" in WB-EMS independently of the setting, the issue of the supervision ratio is an aspect of regular debate. We clearly advocate a maximum supervision ratio of 1 (trainer) to 2 (trainees) in non-medical (Kemmler et al. 2023) and a ratio of 1-1 in medical WB-EMS (Berger et al. 2022).

Not only potential hazards (Strahlenschutzkommission 2019) but in particular ensuring an effective application require a continuous interaction between trainers and trainees. This includes a very close verbal, visual and, if necessary, haptic (tactile) exchange of information. In order to adequately regulate the impulse intensity per electrode during a training session, there must be regular verbal querying of the individual strain (Kemmler et al. 2023), with readjustment of the stimulus intensity (Sect. 3.2.4). A short distance from the trainee also allows the trainer to maintain precise visual control, easy verbal communication, implementation of manual intensity techniques and rapid intervention in the event of an incident or unintended side effect.

> The relevance of consistent physical supervision and guidance of WB-EMS to ensure safety and effectiveness of the application can be considered higher compared with other types of exercise. We therefore advocate a very close supervision ratio.

2.2.10 Training Frequency

Widely independent of the study outcome, the majority of clinical trials apply a low training frequency of 1.5–2 sessions per week (Beier et al. 2024). In commercial, non-medical WB-EMS settings, a single application per week is the standard specification, for which positive effects have been proven for some important issues (e.g. chronic non-specific low back pain (Micke et al. 2021), sarcopenia (Kemmler et al. 2016). Reviewing the reasons for this low training frequency (and volume), we have to stress once again the aspect of time efficiency of WB-EMS as a unique selling point. Additionally, a rationale for the low recommended training frequency was the observation that peak values for biochemical muscle parameters (e.g. creatine kinase) related to severe rhabdomyolysis are observed 3–4 days "post-exercise" (Teschler et al. 2016). This delayed peak after, albeit excessively intensive, WB-EMS in novice applicants suggests that applying further WB-EMS during this period is counterproductive (Teschler et al. 2016). However, the same and other studies (e.g. Hettchen et al. 2019) observed a rather pronounced repeated bout effect on biomarkers related to rhabdomyolysis, thus regeneration periods between the sessions decrease after a few weeks. Nevertheless, so far no study has addressed the important issue of how long a once weekly WB-EMS application remains effective, even when considering the inherent progressive incrementing of impulse intensity during WB-EMS (Sect. 2.2.6).

2.3 Summary of Training and Impulse Parameters Presently Applied in Most Scientific and Commercial WB-EMS Settings

Considering WB-EMS as a resistance type exercise, most of the present scientific and (to a certain extent for training frequency) commercial WB-EMS apply a protocol that captures most of the recommendations listed above, with specific emphasis being placed on low training volume and (after an adequate conditioning period) high impulse intensity.

> Based on a training volume of 1–2 × 20 min per week, a bipolar pulse of 80–85 Hz with an impulse width of 300–400 μs over a duration of 4-6 s with an impulse break of 4 s is used by the vast majority of the present scientific and commercial WB-EMS protocols. The applied impulse intensity of 6–8 Borg CR-10 indicates the high intensity of stimulation. Depending on the training status, aims and impulse duration, predominately direct boosts or a short 0.5 s ramp-shaped rises were applied.

Of note, the WB-EMS protocol described above differs fundamentally from superimposed training used for athletic cohorts/competitive sports at least when addressing discipline specific skills. While emphasis in most scientific and commercial settings is placed on electromyostimulation with adjuvant movements with low or moderate intensity, the approach of superimposed WB-EMS is geared to high intensity to maximum voluntary exercises with added electromyostimulation. Although evidence for significant effects of consistently passive WB-EMS application on functional outcomes has been provided in particular by Belt Electrode-Skeletal Muscle Electrical Stimulation (B-SES),[2] we recommend an active application. This includes joint friendly movements with low intensity and, when needed, low amplitude in order to address functional aspects and skills, which are especially important for older or impaired cohorts.

Of importance however, it should be noted that many of the above specifications and recommendations originate from studies on local EMS that cannot automatically be transferred to WB-EMS (Sect. 1.3). Unfortunately, there is a lack of studies that reliably evaluate and compare the effects of variations in impulse parameters on dedicated outcomes. This includes even popular and frequently addressed study outcomes such as maximum muscle strength, -power, lean body mass or low back pain. This might be due to most of the outcomes (Le et al. 2024) and cohorts (Beier et al. 2024) addressed so far being positively affected by the standard protocol listed above. As a consequence, elaborate and research-intensive comparisons of training and impulse parameters (e.g. Berger et al. 2020; Kemmler et al. 2015; Stephan et al. 2023) have been given far too little consideration in WB-EMS research, a limitation that has to be addressed much more vigor during the nearest future.

> The time efficient, resistance-type approach of most present WB-EMS settings might be at least suboptimal for addressing endurance-affine outcomes. More research should focus on comparisons of relevant training (e.g. training frequency/volume) and impulse (e.g. impulse frequency, intensity) parameters to determine the most effective WB-EMS protocol for a given category of functional and health-related outcomes.

---

[2] I.e. a NMES approach that focuses on gluteals and lower extremity predominately in vulnerable hospitalized people (e.g. Hamada et al. 2023; Homma et al. 2022).

3 Potential Risks, Recommendations and Contraindication for WB-EMS Application

3.1 Potential Risks of WB-EMS Application

Due to the absence of high intensity, voluntary exercise, WB-EMS in general displays a lower orthopedic risk compared to similarly effective high intensity resistance (HIT-RT) exercise (Kemmler, Kohl et al. 2016; Kemmler, Teschler et al. 2016). However, the unique ability of WB-EMS technique to stimulate large areas of the body (up to 2800 cm^2) involving all or at least most major muscle groups with — in excess — supramaximal stimulation intensity may immediately indicate potential cardiometabolic risks of this innovative training technology. In the early years of WB-EMS application, several case studies (e.g. Finsterer and Stollberger 2015; Kastner et al. 2014; Malnick et al. 2016) and public media releases (Habich 2015) reported negative side effects predominately related to severe rhabdomyolysis, observed mainly in WB-EMS novices. With respect to biomarkers of rhabdomyolysis, in excess creatinkinase levels of 240,000 IU, i.e. a 1000-fold increase, have been reported for a highly motivated young soccer player (Kastner et al. 2014). Common symptoms of rhabdomyolysis are severe muscle pain, muscle weakness, muscle oedema, fever, nausea and "cola or tea-colored" urine. Dangerous complications of rhabdomyolysis include acute kidney failure, constriction of the muscle areas with associated circulatory problems and in a worst case cardiac arrest. Considering the pronounced cardiac, hepatic and particularly renal consequences of rhabdomyolysis, these finding were very alarming and have contributed to the official safety warning issued by the Israeli Ministry of Health (Malnick et al. 2016).

A subsequent clinical trial with 26 novice WB-EMS applicants experienced in resistance training (Kemmler et al. 2015; Teschler et al. 2016) that addressed the hardly believable biomarker results listed above confirmed the finding of severe

W. Kemmler et al., *Whole-Body Electromyostimulation*, essentials,
https://doi.org/10.1007/978-3-031-56710-0_3

rhabdomyolysis after an over-intensive WB-EMS first application, however with very pronounced inter-individual variations. Nevertheless, regardless of the CK and myoglobin levels, none of the study participants showed clinical symptoms of rhabdomyolysis. Another observation was that intensity and length of muscle soreness did not correlate with CK levels. Of importance, the study observed a delayed CK and myoglobic peak 3–4 days post-WB-EMS, a feature related to the recommended low training frequency (Sect. 2.2.10).

The underlying reason for NEMS-induced muscular overreach with much higher biomarker levels compared to excentric resistance exercise, for example, (Koch et al. 2014) or long distance running (Teschler et al. 2016) is closely related to the non-voluntary, external stimulation of muscle contraction by the EMS device. In contrast to voluntary muscle stimulation which has to be stopped after muscular fatigue or failure due to energetic undersupply, metabolic acidosis and in particular the inability to recruit all muscle fibers simultaneously, these characteristics of physiologic protection against overreaching do not apply for external NMES stimulus. Thus, despite muscular failure, muscle stimulation was maintained with the consequence of very pronounced physiological, histomorphometric and metabolic muscle responses. Considering the whole-body character of WB-EMS, this unfavorable local effect is multiplied by each muscle group beneath the up to 12 large electrodes. Another aspect of relevance in cases of adverse effects is that unlike conventional exercise, the regulation of exercise (impulse) intensity i.e. the parameter most closely related to the adverse effect, is in the hands of the trainer and not the trainee. In terms of liability law, this aspect is crucial and underscores the high responsibility of the qualified trainer in WB-EMS application.

3.2 Recommendations for Safe and Effective WB-EMS Training

Due to the potential risks of WB-EMS described above, a careful application is required. As a response to Malnick's call "to regulate WB-EMS", (Malnick et al. 2016) a German expert team provided recommendations for safe and effective whole-body electromyostimulation training back in 2016 (Kemmler, Fröhlich et al. 2016). Despite these, as such non-mandatory, specifications, WB-EMS was recently subject to federal regulation in Germany (BMU 2019, 2020). Unfortunately, the proper application of WB-EMS has still to be addressed by these ordinances. In the wake of technical innovations in WB-EMS technology (e.g. wireless WB-EMS), setting (e.g. remote or online WB-EMS) and regulations (e.g. mandatory trainer education) the 2016 guideline has been carefully revised by an international consensus process (Kemmler et al. 2023). The 2023 international recommendations for safe and effective WB-EMS training have been structured into four chapters: (a) "general aspects of WB-EMS", (b) "preparation for training", (c) "WB-EMS application" itself and (d) "safety aspects during and after training" (Kemmler et al. 2023) that will be listed below.

> Due to the rhabdomyolysis-induced enhanced risk profile of WB-EMS application, this exercise technology has been increasingly addressed by recommendations, federal ordinances and contraindications during the last years. As a result, WB-EMS is probably the most extensively regulated training method and can be thus considered as a very safe training technology when applied properly.

3.2.1 General Aspects of WB-EMS Training

1. Comparable to other types of exercise training performed with high intensity, it may well be advisable to have a sports medical examination prior to the WB-EMS training.
2. In order to be safe and effective, WB-EMS training must be provided and supervised by a licensed and, ideally, experienced WB-EMS trainer or, in a university or clinical setting, by scientifically trained staff familiar well versed in its application. Non-supervised WB-EMS application must be strictly avoided.
3. Trainers must have official basic education that qualifies them as coaches according to the laws of their country. In addition to a basic exercise and medical qualification, the licensing process of the trainer should include at least 20 h training on dedicated WB-EMS theory and a practical part provided by an accredited, educationally qualified institution (BMU 2020).
4. We strongly advise a 1:1 trainer-participant ratio (medical WB-EMS), although a 1:2 ratio is also considered acceptable for non-medical WB-EMS applications with less critical participants.
5. Prior to the first WB-EMS session, a detailed anamnesis of possible absolute and relative contraindications (DIN 2019; Kemmler et al. 2019) based on a list of questions must be conducted and documented, confirmed by the

client's signature and archived. While absolute contraindications prevent WB-EMS application in a non-medical WB-EMS setting, a medical practitioner has to give written approval for WB-EMS application in cases of relative contraindications (Sect. 3.3).

6. In parallel, after detailed personal information about WB-EMS, an informed consent contract should be signed by the client/participant to ensure the user understands all the risks and features of the WB-EMS application. We strongly recommend repeating this process at least every six months to update changes in client health, needs or requests that trainers and other responsible persons should take into consideration.

3.2.2 Preparing for Training

1. As with any kind of intensive exercise training, WB-EMS should be only carried out in a proper physical condition and state. This includes abstaining from alcohol consumption, drugs, muscle relaxants or severe stress sufficiently well in advance (i.e., 24–48 h) of the training. WB-EMS is also prohibited when suffering from an illness with fever.
2. WB-EMS training can generate very high metabolic stress because of its derived simultaneous stimulation of all the main muscle groups with high intensity (see below). To prevent weakness, dizziness or other adverse effects related to hypoglycemia during the WB-EMS session, sufficient food intake predominately based on carbohydrates should be ensured in preparation of the session. At least a high carbohydrate, but light snack ($\approx$250 kcal) is recommended, ideally 2 h before the WB-EMS training.
3. In parallel, to minimize renal stress related to intense WB-EMS training, which might be particularly important for individuals with undiagnosed renal problems, additional fluids should be scheduled 30 min before and immediately after training (in each case 250–500 ml or 5 ml/kg body-mass).
4. The trainer has to check the aforementioned aspects before the start of the WB-EMS application by visual inspection and inquiry. Electrode location and proper suit adjustment have to be checked prior to starting the WB-EMS session.

3.2.3 During WB-EMS Application

1. Regardless of the health and exercise status or the participant's ideas and preferences, initial WB-EMS application(s) must be applied carefully. This

especially rules out WB-EMS with high intensity, let alone to exhaustion, during the first 8–10 weeks of WB-EMS application (Teschler et al. 2016).

2. After initial moderate-intensity WB-EMS, i.e. "4" (= somewhat strong) on the Borg CR10 RPE scale (Borg and Borg 2010), the stimulation level or impulse intensity can be subsequently increased and adapted to the individual training aims during the next 7–9 weeks. Intensity levels of "7–8" (= very hard) on Borg CR10 (Borg and Borg 2010) can be allowed after 8–10 sessions of regular training at the earliest. Training to complete exhaustion ("10" at Borg CR10) or continuous tetanus during the impulse phase must be strictly avoided regardless of the training status of the individual.
3. Due to individual differences in impulse sensitivity and tolerance, we recommend using the Borg CR10 RPE scale (Borg and Borg 2010) to prescribe, query and monitor impulse intensity during the WB-EMS session. Of importance, impulse intensity has to be set, queried and monitored several times during the entire session for each individual electrode.
4. Body awareness, individual evaluation and interpretation of the perceived exertion should be a focus of the first exercise sessions.
5. In addition to impulse-intensity issues, the first WB-EMS session should be conducted with a reduced volume. We recommend (a) 5 min impulse familiarization using a continuous WB-EMS protocol and (b) 12 min of intermittent WB-EMS with short impulse phases (4 s), and slow-moderate impulse increases (0.3–0.5 s ramps), intermitted by short breaks (4 s). After 4–6 weeks of familiarization/conditioning the WB-EMS session can be carefully increased to a maximum of 20 min applying an intermittent resistance exercise training (RT)-type protocol with high impulse intensities (see above) or 30–40 min session applying an endurance typeprotocol that was further recommended to consistently schedule (only) moderate impulse intensity (5–6, i.e. "hard" to "hard+" on Borg CR10).
6. Further, for adequate recovery and adaptation and to prevent potential health impairments, the training frequency may not exceed one training session of 20 min per week during the initial 8–10 weeks.
7. After this 8–10-week familiarization and conditioning period, there must be a 4-day break between intense WB-EMS sessions ($\geq$7 Borg CR10) to avoid accumulation of muscle breakdown products and permit adequate regeneration and adaptation.

3.2.4 Safety Aspects during and after Training

1. During the WB-EMS session, the trainer has to exclusively focus on the well-being of the clients/participants. Before, during and after the training session, the trainer has to verbally and visually check the participant's condition so as to rule out health risks and ensure effective training. The training session must be stopped immediately in case of any adverse effects or problems being observed by the trainer or raised by the participants.
2. We strongly recommend a very close interaction and proximity between trainer and participant and maintaining a specific focus on the following key points: (a) frequent feedback about perceived exertion for each area of stimulation, (b) permanent visual monitoring of the participant and eye contact to check participant strain, avoid overload and to react immediately to the first signs of adverse effects, and (c) verbal and haptic movement corrections and rapid assistance in cases of emergency.
3. With respect to the frequent request for higher impulse intensity mentioned above, we suggest checking the adequacy of impulse intensity at least 3 times per area/electrode by verbal query. In cases of inadequate impulse intensity, levels should be adjusted in close interaction between trainer and participants to ensure a safe and effective WB-EMS application.
4. Operating controls must be constantly in reach of the trainer and the participant (i.e. maximum distance of 120 cm (Deutsches Institut für Normung 2020)) in order to stop the WB-EMS application immediately in case of emergency. The participant has to be briefed on the emergency shutdown function of the device.
5. Medical consultation and clarification is advisable in the case of relevant discomfort, cardiometabolic difficulties or orthopedic problems potentially related to the WB-EMS application. This also refers to hematuria (e.g., cola-colored urine), persistent headache and inflammatory or bleeding problems after potentially too intense WB-EMS application.

> Although non-mandatory but applied by the majority of commercial WB-EMS providers, the revised international recommendations can be considered as a cornerstone of safe and effective WB-EMS. The high acceptance of the recommendation might reflect the applicable concept and the consideration of stakeholder input during the consensus process.

3.3 Contraindications for the Use of Non-Medical WB-EMS

Besides the recommendations for safe and effective WB-EMS training discussed above, another safety issue is the identification of cohorts with particular risk for WB-EMS application. In this context, "the recommended contraindications for the use of non-medical WB-Electromyostimulation" was released by a German expert group in 2019 (Kemmler et al. 2019). The limited regulation of WB-EMS, non-mandatory instructor education and evidence gaps on conditions and diseases considered particularly critical for WB-EMS application led to a very restrictive list of absolute and relative contraindication in 2019. The recent federal regulations, mandatory trainer education, revised recommendations and new scientific evidence for WB-EMS applications in vulnerable cohorts justified the careful revision of the contraindications in 2023/2024 by a multidisciplinary consortium. The final decision on contraindications was based on the number of studies that addressed the cohort in question, quality criteria of the trial and safety aspects with specific regard for adverse effects related to the WB-EMS intervention. The decisive factor was ultimately the benefit/risk assessment of the criteria. Contraindications were categorized into relative and absolute contraindications. In the case of absolute contraindications, WB-EMS training should be ruled out due to the significant potential danger or harm. The main guideline here is that the event is of a significantly health-impairing nature or might cause a life-threatening situation. In contrast, relative contraindications mean that the benefit/risk ratio should be carefully considered by a medical practitioner who decides on the approval.

> Due to fixed electrodes in vests and belts, some contraindications of local electromyostimulation (e.g. transcranial application) are not applicable for WB-EMS; however, most contraindications of local EMS (review in van Kerkhof 2022) do also apply to WB-EMS applications.

3.3.1 Absolute Contraindications for the Use of Non-Medical WB-EMS

Acute illnesses, bacterial infections and inflammatory processes: Regardless of the acute restriction caused by the illness, the body is subject to increased immunological stress after exercise. This significantly weakens the body and makes it more susceptible to further infections, which is why exercise or WB-EMS training is generally strongly discouraged.

Recently performed operations in stimulation areas: Operations involving open or sutured wounds prevent WB-EMS training if the wound is covered by, or closely located to the electrode. In addition, the original surgical site should have recovered completely.

Arteriosclerosis, arterial circulatory disorders: Arteriosclerosis is characterized by pathological deposits of blood lipids (plaques) on the inner wall of arterial vessels. Although several studies addressed cohorts with atherosclerosis, arterial circulation disorders and related diseases (Beier et al. 2024) without reporting any adverse effects of WB-EMS application, potential events can be life-threatening, thus non-medical WB-EMS training is absolutely contraindicated.

Stents and bypasses that have been active for less than 6 months: In the case of stents and bypasses, the corresponding heart surgery can involve a serious intervention in the human organism. As WB-EMS training is a high-intensity exercise, it is essential to avoid such exercise during post-operative rehabilitation in the first six months after the operation.

Untreated hypertension: Treated and controlled hypertension does not impair the ability to exercise at all. However, as untreated high blood pressure is associated with an increased risk of stroke and heart attack as well as renal insufficiency, a medical check-up should be carried out first and WB-EMS training should be avoided.

Pregnancy: In contrast to documented training recommendations and contraindicators of general physical training during pregnancy and postpartum (Sulprizio and Kleinert 2016), no evidence is provided for WB-EMS application during these periods. To avoid possible damage to the fetus or the triggering of premature labor, we advise against whole-body EMS during pregnancy. Whether WB-EMS training can be applied in the early post-pregnancy period cannot be answered reliably.

Electric implants, cardiac pacemakers: WB-EMS training applies electrical impulses of different frequencies, width, raises and intensities. To date, there is a lack of manufacturer information on possible interference of the applied impulse of WB-EMS training with electrical implants. As a negative influence cannot therefore be ruled out per se, we advise against any kind of WB-EMS.

Heart arrhythmia: Does not fundamentally rule out a low-moderate intensity conventional type of exercise. However, no robust evidence has yet been provided for WB-EMS training. Apart from the aspects of external impulses, similar to other types of high intensity exercise, WB-EMS may not be used in cohorts with heart arrhythmia due to the potentially life-threatening consequences.

Severe bleeding disorders, tendency of bleeding (hemophilia): Severe and life-threatening bleeding disorders should be considered as an absolute contraindication for WB-EMS. Although there is some evidence for functional effects of local EMS in patients with bleeding disorders (e.g. Querol et al. 2006), the authors focus on an isolated muscle group under laboratory conditions with constant supervision by medical staff. However, due to the limited evidence and high risk in case of events, we consider this category as an absolute contraindication for non-medical WB-EMS application.

Abdominal wall and inguinal hernia: Dependent on their size, abdominal wall and inguinal hernia represent serious injuries in the abdominal area. Physical strain or corresponding tensile or compressive stress can result in an enlargement of the hernia, associated leakage or even injury to internal organs. For this reason, hernias have to be medically treated and fundamentally exclude physical training and especially WB-EMS training with electrodes in the area of the hernias.

Acute influence of alcohol, drugs and intoxicants: General physical training and WB-EMS training under the influence of alcohol, drugs, psychotropic drugs or intoxicants in various forms and levels must be ruled out as a matter of principle due to possible danger and harm. This particularly refers to WB-EMS with its pronounced demands on body awareness and sensitivity.

> Due to the cardiometabolic risk profile, high regulatory focus by federal authorities and the enhanced responsibility of the staff, the absolute contraindications on WB-EMS application for the non-medical setting can be considered as very cautious. One background to these restrictive, albeit still non-mandatory, contraindications is to prevent further drastic regulation by federal authorities.

3.3.2 Relative Contraindications for the Use of Non-Medical WB-EMS

In contrast to absolute contraindications, relative contraindications are restrictions that require medical approval before WB-EMS sessions can be undertaken.

- Diabetes mellitus (Type I and II)
- Tumor and cancer
- Acute back pain without diagnosis
- Acute neuralgia, herniated discs

- Implants older than 6 months
- Diseases of the internal organs particularly kidney diseases
- Cardiovascular diseases
- Movement kinetosis
- Greater fluid retention, oedema
- Open skin injuries, wounds, eczema, burns (in the vicinity of electrodes)
- Corresponding medication for conditions mentioned above

Based on recent scientific research and benefit/risk assessment, Diabetes mellitus (Type I and II) and Tumor/Cancer have been moved from the absolute to the relative contraindication catalogue. Apart from the catalogue displayed above, symptoms whose cause is initially unknown without medical clarification are also classified as relative contraindications. As an example, the causes of movement kinetosis or edema can be harmless, but can also be the main symptom of a previously undiagnosed serious disease. It is within neither the competence nor the responsibility of EMS trainers to properly diagnose these complaints. In this respect, medical clarification is mandatory in these cases before the WB-EMS application. The risk assessment for diseases of internal organs or implants also exceeds the scope of competence of EMS trainers. In these cases, too, medical clarification is mandatory prior to the application of WB-EMS.

Open skin injuries or severe extensive skin irritations (e.g. allergies, eczema, active neurodermatitis, etc.) in the area of the electrodes also count as potential exclusion criteria for WB-EMS. In the presence of sunburn, trainers have to clarify the severity and area of the sunburn with the users and only then decide to what extent a WB-EMS application is feasible.

We have not revised the list of relative contraindications, so as to leave the physician as the gatekeeper of the process. In this context, one may criticize that most physicians might be unable to estimate the risk and benefits of WB-EMS well enough to allow a WB-EMS application. We do not agree with this. Considering the commercial application since 2007 with thousands of studios, millions of clients and hundreds of publications in medical journals (Beier et al. 2024), most physicians are well aware of WB-EMS.

> Even this updated list of contraindications should not be regarded as the final version. In view of the greatly increased number of WB-EMS trials with a majority of the studies that focus on cohorts with limitations or diseases and less intensive WB-EMS protocols, the absolute contraindications may need updating in the next years.

Evidence of WB-EMS on Different Outcomes

4

4.1 Health-Related Outcomes

Health-related parameters are key outcomes of many WB-EMS trials (Le et al. 2024). In parallel, many studies focus on participants with diseases, conditions or other health related limitations (Beier et al. 2024). Considering the joint-friendly, time efficient, highly customizable and rigorously supervised setting of WB-EMS, WB-EMS might be one of the few options for effective and safe disease prevention and/or therapy even for (very) vulnerable cohorts otherwise unable or unmotivated to perform frequent and intense exercise protocols. Applying the resistance-type based WB-EMS provided by most scientific and mainly all commercial settings (Sect. 2.3), the effects on health parameters are roughly similar to those of low-volume, high-intensity resistance exercise training (HIT-RT) (Kemmler, Kohl, et al. 2016). However, in contrast to RT, which has been the subject of intensive research for many decades, significantly less evidence for positive effects is available for WB-EMS. However, two evidence maps that addressed cohort (Beier et al. 2024) or outcomes (Le et al. 2024) covered by WB-EMS trials (i.e. longitudinal studies) indicate intensive scientific work on WB-EMS applications particularly in the field of health-related outcomes and/or cohorts with diseases (Figs. 4.1, 4.2 and 4.3).

4.1.1 Non-Specific Chronic Low Back Pain

Evidence of positive WB-EMS effects on non-specific chronic low back pain (NCLBP) is largely confirmed. Applying standard WB-EMS protocols (Sect. 2.3) several research projects (Kemmler et al. 2017; Konrad et al. 2021; Micke et al.

W. Kemmler et al., *Whole-Body Electromyostimulation*, essentials,
https://doi.org/10.1007/978-3-031-56710-0_4

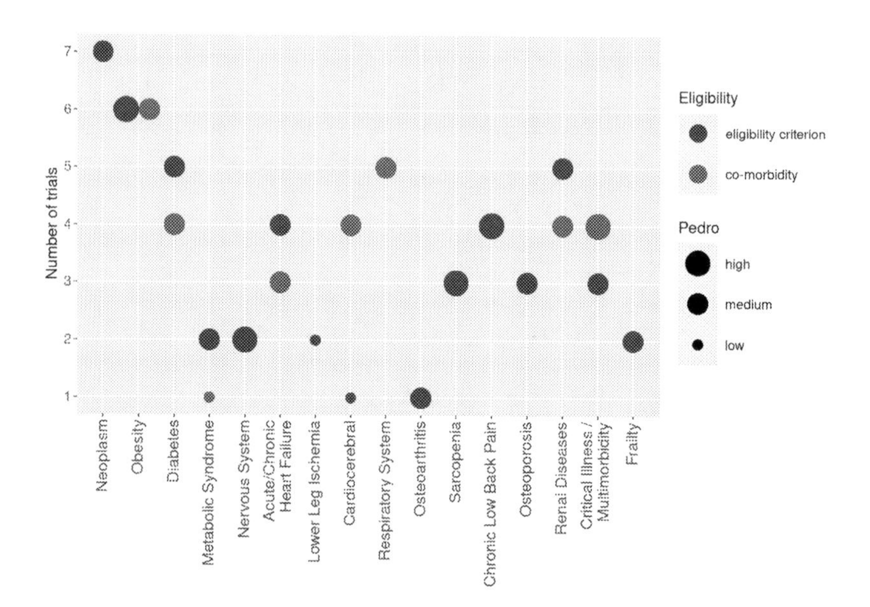

Fig. 4.1 Cohorts with diseases and conditions addressed by WB-EMS. The y-axis represents the number of studies that focus on the corresponding cohort (x-axis). Different colors indicate whether the health status of the cohort was applied as a criterion for inclusion (blue) or reported as a simple co-morbidity (green). The size of the bubble indicates the methodological quality according to PEDro. The biggest size indicates at least one study of high methodological quality in the domain. The lowest size of the bubble chart represents at least one study of low methodological quality in the domain

2021; Silvestri et al. 2023; Weissenfels et al. 2018) observed significant positive effects on NCLBP as the primary study outcome. Additionally, two trials (Konrad et al. 2021; Silvestri et al. 2023) reported favorable effects on the Oswestry Disability Index (ODI) that focuses on low back pain induced (dis)ability to manage everyday life. Studies that compared WB-EMS with usual care protocols reported similar effects to dedicated resistance exercise or whole body vibration programs (Micke et al. 2021). Particularly impressive, Konrad et al. (Konrad et al. 2020) reported significantly more favorable effects of once weekly WB-EMS on pain intensity and ODI compared with a multimodal low back pain program four days per week in an outpatient setting.

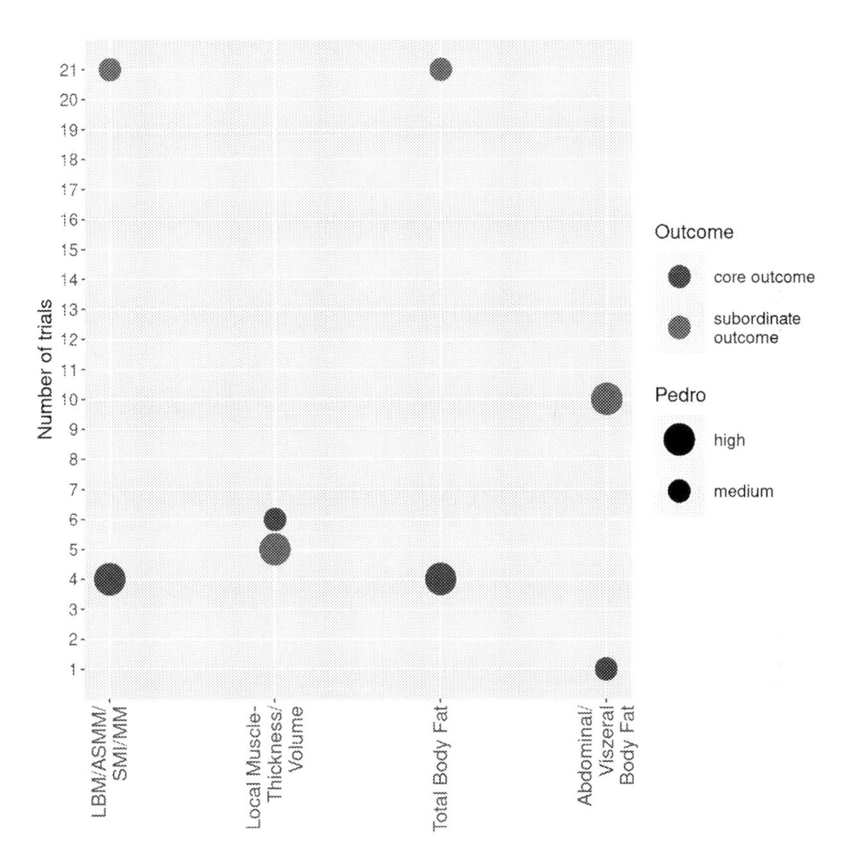

Fig. 4.2 Body composition-related outcome addressed by WB-EMS and classified according to the aspect whether the domain was considered as a core (primary) or subordinate outcome

- There is robust evidence for positive effects of intermitted low volume, moderate-high intensity WB-EMS protocols on chronic non-specific low-back pain parameters.

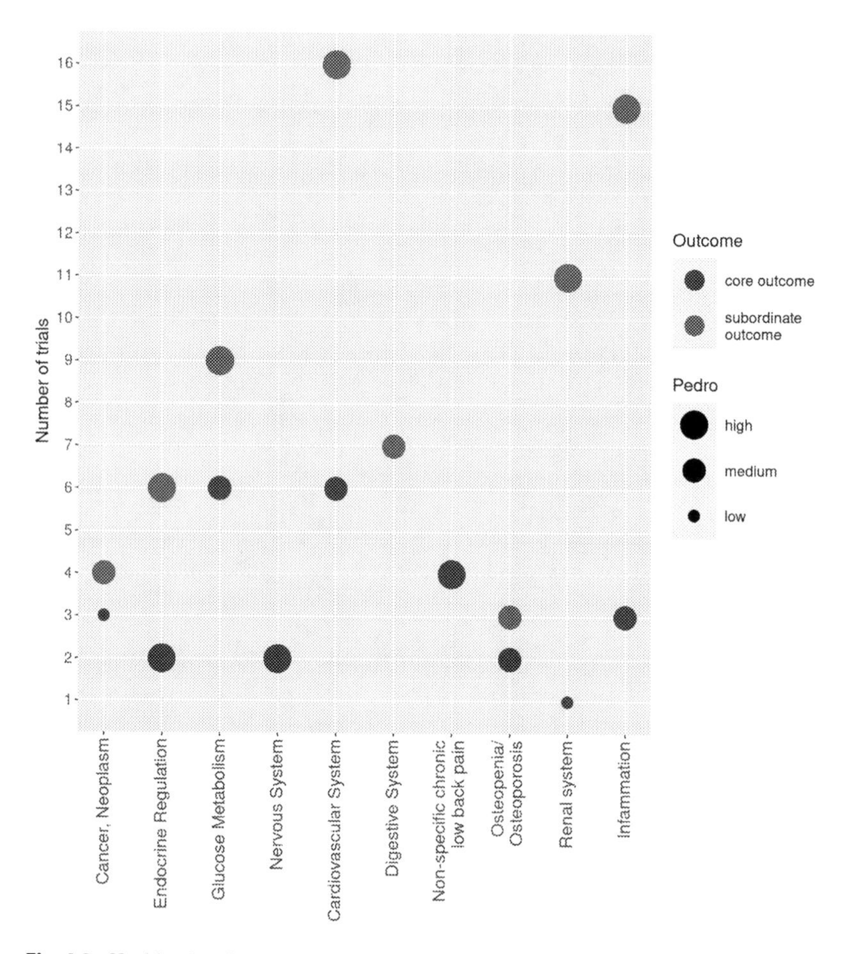

Fig. 4.3 Health-related outcome addressed by WB-EMS and classified according to the aspect whether the domain was considered as a core (primary) or subordinate outcome

4.1.2 Sarcopenia, Muscle Wasting

Sarcopenia, the combined reduction of muscle mass and function[1] was considered as a primary outcome by two studies (Kemmler, Teschler et al. 2016; Kemmler

[1] WB-EMS effects on muscle function were discussed in chapter (Sect. 4.3).

et al. 2017). At least three more studies (Kemmler et al. 2014; Kemmler and von Stengel 2012; Tsurumi et al. 2022) addressed dedicated morphometric (i.e. appendicular muscle mass, skeletal muscle mass index) and functional parameters (strength, power, muscle function) involved in the complex sarcopenia diagnosis (Cruz-Jentoft et al. 2010, 2019). In all cases, WB-EMS intervention showed clinically relevant effects on sarcopenia variables. Besides favorable changes of muscle mass, cross sectional area/volume, muscle thickness and other proxies of hypertrophic effects, one study in older men with sarcopenic obesity (SO) (Kemmler et al. 2018) reported significant positive effects after standard WB-EMS (and protein supplementation) on skeletal muscle fatty infiltration and cross sectional muscle volume. Lastly, apart from effectiveness and safety (Kemmler et al. 2020), the joint-friendly, consistently supervised WB-EMS might be particular suitable and attractive for the rather vulnerable sarcopenic cohorts.

> Due to its individualized, consistently supervised and joint-friendly nature, WB-EMS is particularly attractive for the resistance-oriented training on Sarcopenia in older people. Apart from physical exercise, be it WB-EMS or conventional resistance exercise, the protein status of participants should be also considered for generating favorable effects on muscle mass parameters.

4.1.3 Osteopenia and Osteoporosis

Due to the close interaction between muscle and bone, WB-EMS, with its considerable effect on muscle parameters, should also trigger relevant effects on bone strength. Bone mineral density (BMD) or bone mineral content (BMC), i.e. outcomes specifically linked to osteopenia or osteoporosis, have been determined in five studies (Amaro-Gahete et al. 2019; Müllerová et al. 2022; Sánchez-Infante et al. 2020; Vaculikova 2022; von Stengel et al. 2015) so far. However, only one study provided more than six months of WB-EMS application (von Stengel et al. 2015), i.e. a period that exceeds the bone remodeling cycle (Eriksen 2010). In summary, this 12-month study determined the effect of standard WB-EMS on BMD in women 70 years and older with diagnosed osteopenia. Briefly, the study showed borderline significant effects for the lumbar spine but negligible effects for the femoral neck BMD. Two reasons might have contributed to this suboptimum result. Apart from the lack of impact loading by standard WB-EMS application, there is considerable evidence that training frequencies below 2 sessions per week are ineffective for increasing BMD in adults (Zitzmann et al. 2022). However, there is no rationale against the implementation of a few impact exercises conducted during standard WB-EMS. In conclusion, non-adapted standard WB-EMS protocols are only moderately effective in increasing BMD in older adults.

> Stimuli triggered by present standard WB-EMS protocols are not perfectly suited for affecting BMD. When addressing BMD as the main outcome of the WB-EMS application, minor adjustments of the training protocol should be considered.

4.1.4 Osteoarthrosis

So far only one study (Park, Min et al. 2021) has determined the effect of a superimposed, intermitted (6 s–4 s), low frequency WB-EMS applied 3 × 20 min/week on knee osteoarthritis in elderly women with early knee osteoarthritis (OA). Outcomes of the study included knee pain and symptoms, ADL summarized by the KOOS Score (Roos and Lohmander 2003) and inflammatory biomarkers. In summary, the authors reported significant positive effects on pain, symptoms, ADL, and Interleukin 6 compared with an active CG, which conducted the same isometric exercises but without superimposed WB-EMS. Another presently unpublished study (Kast et al.) that applied a 7-month standard WB-EMS protocol (Sect. 2.3) in overweight to obese people with advanced OA of the knee reported significant effects on all KOOS categories including pain, symptoms and ADL, verified by a 7-day knee pain protocol conducted at baseline and follow-up. One may argue that local NMES stimulation might be less complex but similarly effective for the (local) stimulation of knee OA, for example. However, simultaneous activation of agonists and antagonists of the thigh using belt electrodes ensures harmonious strengthening of the quadriceps and ischiocrural muscles, which is beneficial for joint stability. Equally important, there is some evidence that WB-EMS decreases visceral fat (Jee 2019), which is involved in the development and progression of knee OA.

> Due to its joint-friendly character WB-EMS is a suitable and attractive training technology for people with OA. Despite the regionally limited OA, WB-EMS is preferable to local application due to its systemic effect.

4.1.5 Obesity, Severe Overweight

At least twelve studies (Beier et al. 2024) addressed cohorts with obesity or sarcopenic obesity (SO) while applying different WB-EMS protocols. Correspondingly, results on total and regional body fat changes differ considerably. Applying a once weekly low intensity standard WB-EMS protocol Kemmler et al. (Kemmler, Teschler et al. 2016) failed to determine significant effects on total and abdominal body fat in older women with SO. On the other hand, 1.5×20 min of moderate-high intensity WB-EMS resulted in significant effects on total and regional fat changes in older men with SO (Kemmler, Weissenfels et al. 2017). Most impressively, Kim et al. (Kim and Jee 2020b) reported significant effects on body fat reduction of 4.5 kg, accompanied by muscle mass increases, after 8 week of albeit 3×40 min superimposed (aerobic dance) standard WB-EMS compared to a CG provided with 4×40 min of aerobic dance without WB-EMS. WB-EMS trials that evaluated the effect of WB-EMS under energy restriction (500–600 kcal per day) in obese people showed no significant additive effects on body fat parameters compared to diet alone (Bellia et al. 2020; Reljic et al. 2021; Willert et al. 2019). Reviewing the evidence, most WB-EMS studies observed positive changes on body fat parameters, including visceral adipose tissue (Jee 2019); however the effects were not always significant. Briefly reviewing the mechanisms of WB-EMS effects on body fat changes induced by increments of energy expenditure, at least three effects can be listed.

1. The acute energy expenditure during WB-EMS, which is limited by the low training volume of standard WB-EMS.
2. The post-exercise effect induced by energy restoration, repair and adaptive processes post-exercise (EPOC) that is particularly pronounced after WB-EMS application (Teschler et al. 2018).
3. Changes in basal metabolic rate due to the hypertrophic effects of WB-EMS (Sect. 4.2.1).

> WB-EMS application results in positive effects on overweight and obesity, however the effect size of isolated standard WB-EMS protocols can be considered moderate at best. The relevance of WB-EMS in the field of severe overweight and obesity is much more to prevent the pronounced reduction of muscle mass by most diets that focus on energy restriction.

4.1.6 Non-Insulin Dependent Diabetes Mellitus (NIDDM, Diabetes Mellitus Type II) and the Metabolic Syndrome

Until recently, NIDDM was considered as an absolute contraindication for commercial, non-medical WB-EMS (Sect. 3.3.1). Although this limitation does not apply to WB-EMS research, it might have caused the paucity of research in this area. Despite the many studies that addressed biomarkers related to glucose metabolism, only two studies[2] applied WB-EMS in cohorts with NIDDM (Houdijk et al. 2022; van Buuren et al. 2015). While van Buuren et al. reported significant improvements in glucose metabolism (e.g. $HbA1_c$) and aerobic capacity (van Buuren et al. 2015), with the same 2×20 min standard WB-EMS, Houdijk et al. (2022) reported significant positive effects on $HbA1_c$ only in his male patients. A further three RCTs that focused on people with the Metabolic Syndrome (Bellia et al. 2020; Kemmler et al. 2010; Reljic et al. 2020) confirmed the favorable effects of (standard) WB-EMS on outcomes related to insulin resistance, glucose metabolism and lipid profile. Another advantage of WB-EMS in the usually less exercise affine cohort of NIDDM patient is its high attractiveness indicated by low drop-out and high attendance rates. Of importance, due to limited sensitivity and corresponding problems with impulse intensity specification, patients with diabetic polyneuropathy should undergo exercise very cautiously in a closely medically supervised setting.

> WB-EMS can be considered as an effective, feasible and attractive component in the therapy of non-insulin dependent Diabetes Mellitus. Of importance so far, no WB-EMS study has addressed cohorts with Insulin dependent diabetes mellitus (DM Type I), which is another cohort that might be attracted by time-efficient training methods.

[2] Additionally three studies on Belt Electrode-Skeletal Muscle Electrical Stimulation (B-SES) (review Beier et al. 2024) confirmed the positive effects of WB-EMS in patients with NIDDM.

4.1.7 Cancer/Tumor/Neoplasms

Recently (Sect. 3.3.1) cancer/tumor/neoplasms were moved from an absolute to a relative contraindication for commercial, non-medical WB-EMS. This progress is very welcome since WB-EMS in a carefully supervised setting represents a feasible and attractive training option for the vulnerable and physically limited cohort of advanced cancer patients. In summary, seven WB-EMS studies addressed cohorts with malignant neoplasms. In their ongoing "advanced cancer" project, Zopf et al. (e.g. Schink et al. 2018; Schwappacher et al. 2021) have so far published study results on cohorts with hematological malignancies, gastro-intestinal, pancreatic, prostate and colorectal cancer. Of importance, only a few trials focus on biomarkers or other outcomes directly related to cancer/tumor/neoplasms development (Schwappacher et al. 2020, 2021). The latter studies reported inhibited growth of human cancer cell cultures in vitro and simultaneously enhanced cell death of tumor cells in blood serum collected from WB-EMS applicants. More studies focus on cachexia and fatigue syndrome in cancer patients. Summarizing these studies (e.g. Schink et al. 2020), there is strong evidence for positive effects of WB-EMS application added to a protein- and energy-enriched diet in stabilizing or increasing muscle mass and function and positively impacting the fatigue syndrome in advanced cancer patients.

> WB-EMS added to protein- and energy-enriched diet can be considered as an effective, feasible and attractive component in the therapy of (advanced) cancer patients. Nevertheless, similar to NIDDM a careful WB-EMS application under supervision of instructors with disease-specific competence is advised.

4.1.8 Cardiovascular and Cerebrovascular Diseases

A large variety of WB-EMS studies focus on risk factors related to hypertension, dyslipidemia or cardiorespiratory fitness, but not predominately on cohorts with such diseases. With respect to hypertension, some studies (e.g. Bellia et al. 2020; Wittmann et al. 2016) reported significant reductions of blood pressure in participants with the Metabolic Syndrome and/or Obesity, whilst others applying the same protocol in comparable cohorts failed to determine significant effects (e.g. Kemmler and von Stengel 2012; Reljic et al. 2020). However, since most cohorts addressed by WB-EMS studies had been on average normotensive, the relevance of this data is negligible. Only few studies determined the effects of acute WB-EMS on blood pressure in hypertensive people (e.g. Kemmler et al. 2024). Widely independent of impulse intensity the latter study reported no relevant increases of systolic or diastolic blood pressure during a standard WB-EMS session in novice applicants. Nevertheless, in summary, more dedicated studies should address these issues in the nearest future.

Reviewing the effect of WB-EMS as a therapy for cardiovascular or cerebrovascular diseases, three studies were eligible. Two studies focus on patients with chronic heart failure applying standard WB-EMS protocols (Fritzsche et al. 2010; van Buuren et al. 2013) twice a week in a clinical setting. Both studies reported significant effects on cardiorespiratory fitness, van Buuren et al. (2013) further observed significant positive effects on left ventricular function compared with a locally applied (gluteals and legs) NMES protocol. Applying an unusual WB-EMS protocol (25 kHz, 5 ms, 150 mA) Lukashevic et al. (Lukashevich 2020) determined significant positive effects on functional parameters and stroke indices in patients<6 months after the stroke event.

> Due to the unspecific inclusion criteria, it is unclear whether there is a clinically relevant effect of WB-EMS on hypertension and dyslipidemia. Although limited evidence prevents a final recommendation, WB-EMS might be also a training option for treating cardiovascular or cerebrovascular diseases. Due to the risk of harmful adverse effects, the WB-EMS application should be restricted to medical settings.

4.1.9 Diseases of the Nervous System

Only one study, di Cagno et al. (2023), compared a 2 × 20 min/week standard WB-EMS (85 Hz, 350 µs, 4 s/4 s) superimposed on isometric exercises vs. a continuous low frequency WB-EMS protocol (7 Hz, 350 µs) superimposed on rowing exercise vs. an inactive control in people with mild to moderate Parkinsons disease (PD). In summary, the endurance-type WB-EMS resulted in significantly more favorable results for PD specific serum biomarkers, fatigue, walking ability and balance compared to the standard protocol (...or control). The authors concluded that endurance-type WB-EMS is a feasible and effective method for reducing the duration of weekly exercise in PD patients.

4.2 WB-EMS Effects on Body Composition

4.2.1 WB-EMS and Hypertrophic Effects

A large number of studies (Le et al. 2024) determined the effect of WB-EMS on lean body mass (LBM) and related parameters in different cohorts, with diverging assessments and varying research questions. Overall, with a few exceptions (e.g. energy restriction (Reljic et al. 2022), bariatric surgery (Ricci et al. 2020), the vast majority of studies consistently revealed (significant) positive hypertrophic effects. Summarizing these effects on muscle mass parameters, a recent meta-analysis (Kemmler et al. 2021) reported a high effect size (SMD: 1.23; 95%-CI: 0.71–1.76) after WB-EMS application.

> In summary, positive evidence for significant WB-EMS effects on lean body mass and related parameters is very high. The size of the hypertrophic effects is roughly similar to HIT-RT (Kemmler, Teschler et al. 2016).

4.2.2 WB-EMS and Total and Regional Body Fat Changes

More than half of the present WB-EMS trials address total and/or regional (predominately trunk and total abdominal, rarely visceral) fat mass as a primary or secondary study outcome. In contrast to hypertrophic effects, study results on body fat changes vary considerably from highly significant effects on body fat reductions (Kim and Jee 2020a) to negative effects (Jee 2019). This disappointing result might be due to baseline participant characteristics and specification of the WB-EMS protocol with regard to frequency and particularly (impulse) intensity. Of note, even frequent WB-EMS applications (3 × 20 min/week) do not necessarily ensure positive effects on body fat mass in normoweight cohorts—at least when applying low or moderate impulse intensity (Jee 2019). Unfortunately, only a few WB-EMS studies properly monitor dietary intake as the key confounding factor, thus, it is a daunting task to reliably summarize WB-EMS effects on body fat. As discussed above (Sect. 4.1.5), the additive effect of WB-EMS to energy restriction on total or regional body fat reduction is limited. More importantly, WB-EMS (and compensatory protein supplementation) prevents the pronounced loss of muscle mass during energy restriction (Sect. 4.1.5) and thus maintains basal metabolic rate (BMR) as a key aspect of energy turnover.

> Similar to resistance exercise, WB-EMS triggers short, moderate and long-term changes of energy turnover. However, due to limited acute effects and low training volume WB-EMS is not the perfect tool for losing total and abdominal body fat. But thanks to the maintenance of the basal metabolic rate WB-EMS can play a vital role in the long term management of overweight and obesity.

4.3 Effects of WB-EMS on Physical Fitness and Functional Outcomes

4.3.1 Strength, Power, and Resistance Related Outcomes

Properly applied standard WB-EMS (Sect. 2.3), i.e. WB-EMS designed as a resistance type protocol, undoubtedly provides significant and clinically relevant effects on outcomes related to maximum strength (≥30 studies), power (≥15 studies) and strength endurance (≥10 studies). Reviewing the large number of studies that focus on strength-related outcomes, most studies applied isometric tests or functional tests while only a few studies used advanced technologies (e.g. isokinetic devices). Unfortunately, the large variety of different test procedures make it hard to quantify effects by meta-analysis. Although this might not be the key argument in favor of conducting WB-EMS or better conventional resistance exercise, (RT) a frequent issue is the difference in their effectiveness on strength-related parameters. However, there are rather limited data on this issue. One study that compared standard WB-EMS vs. HIT-RT (Kemmler, Teschler, Weissenfels, Bebenek, Fröhlich, et al. 2016) reported non-significantly lower positive effects on tests applied in a dynamic mode (i.e. bench press) but non-significantly higher effect on isometric testing (i.e. back extensor) after WB-EMS vs HIT-RT.

> There is some evidence that standard WB-EMS application yields roughly similar effects on maximum strength parameters in untrained cohorts – at least compared with HIT-RT.

Another question concerns superimposed WB-EMS (Sect. 2.2.8), a method predominately applied with athletes. As discussed above, the main difference to "isolated" WB-EMS is the high intensity of voluntary exercise superimposed by WB-EMS that still allows correct execution of the discipline-specific exercise.

Although there is some evidence for tendentially higher effects after superimposed versus isolated exercise (Ludwig et al. 2020), results predominately failed to reach statistical significance (Ludwig et al. 2020; Wirtz et al. 2019). Having said that, the ability to statistically confirm small but nevertheless relevant effects in athletic performance is rather limited in any case considering the low-moderate sample size even provided by dedicated meta-analyses (Wirtz et al. 2019).

> Most studies with athletes reported low, non-significant effects of exercise superimposed by WB-EMS (vs. non-superimposed control group). However, considering the relevance of small, statistically hardly recordable improvements in performance in highly trained athletes, it would be wrong to discourage the application of WB-EMS.

4.3.2 Functional Outcomes in Older Adults

As discussed above, we consider standard WB-EMS as an intervention that is particularly suitable and attractive for older cohorts unmotivated or unable to conduct conventional exercise protocols. Thus, a brief paragraph will particularly address WB-EMS induced mobility-, agility- and strength-related outcomes in this cohort. The majority of WB-EMS studies (n > 10) provided positive effects on mobility and or agility variables such as habitual walking speed, or (less consistently) the timed up and go test, though there is a close correlation to the baseline physical status of the participants and the intensity of the WB-EMS application. WB-EMS studies that recorded parameters in the lower extremities consistently displayed significant and clinically relevant effects, particularly for hip and knee extensor strength (e.g. Kemmler et al. 2018), muscle groups particularly relevant for mobility, morbidity and mortality of older people (Visser et al. 2005). Of importance, even in a supine position, slight movements during WB-EMS demonstrate significantly better results compared with passive application (Kemmler et al. 2015). Handgrip strength, also a highly relevant functional ability in everyday life of older cohorts, was rarely addressed by WB-EMS trials (n = 4). In summary, all the studies reported positive effects that did not consistently reach significance, however (Kemmler, Teschler et al. 2016). One trial that focused on WB-EMS application in assisted living facilities (Bloeckl et al. 2022) might be a blueprint for the promising application of WB-EMS to increase functional ability in frail cohorts. In this study, Blöckl et al. (2022) reported significant positive changes of the short physical performance battery (SPBB), maximum dynamic hip/knee extension/flexion strength, handgrip strength and choice stepping reaction time. The recent study of Park et al. (2023) with pre-frail women widely confirmed the positive WB-EMS effects on functional parameters in older, functionally limited cohorts. Apart from Blöckl et al. (2022) and Park et al. (2023; via SPBB), few other WB-EMS studies determine balance ability in older cohorts (di Cagno et al. 2023; Pano-Rodriguez et al. 2020).[3] However, including results of the more frequently applied B-SES application[4] considerably increases the evidence for positive effects of WB-EMS/B-SES on balance performance (e.g. Homma et al. 2022; Noguchi et al. 2018). A literature review by Paillard et al. (Paillard 2021) largely based on local EMS/NMES concluded that balance performance can be significantly improved in frail/elderly people, while the effects in young/healthy subjects are negligible to weak. Overall, the authors stated that EMS effects are less pronounced compared to voluntary muscle activation. This might be due to the limited use of sensory information for the central nervous system because of the simultaneous activation of sensory and motor neurons.

> Older, functionally limited people might benefit most from WB-EMS, due to improvements in functional parameters closely related to mobility, independence and morbidity.

4.3.3 Endurance Performance

Briefly, endurance is defined as the physical capability to sustain an exercise for an extended period, but also the ability of rapid regeneration. The latter aspect was addressed by a WB-EMS study (1 Hz, 350 ms, 20 min continuous application) by de la Cámara Serrano et al. (2018) that reported similar effects compared with conventional active or passive recovery methods (in line with results on local NMES (Pinar et al. 2012)). Many more studies ($n \geq 19$) provide evidence for performance parameters and physiologic markers of endurance capacity (e.g. VO_2max). Most of the studies applied the 6 min walking test (6MWT), spiroergometry data have been reported by eight studies so far. However, due to the

[3] The study of di Cagno et al. (2023) focused on people with Parkinsons disease, thus the relevance of this study for WB-EMS-induced balance improvements in older people is limited.

[4] With respect to "balance ", it might be justified to include a NMES technology restricted to gluteals and lower extremities.

varying study protocols (e.g. RT- or endurance-type WB-EMS, superimposed or not, different training frequency, active vs. inactive CG) results vary considerably. Studies that applied standard WB-EMS protocols (Sect. 2.3) without relevant additional exercise reported significant effects (6MWT) compared to widely inactive CG only for vulnerable cohorts (e.g. di Cagno et al. 2023; Ricci et al. 2020; Schink et al. 2020; Teschler et al. 2021), while healthy older people do not usually benefit (e.g. Vaculikova 2023). Comparing endurance protocols superimposed by WB-EMS, be it with intermitted or continuous impulse, with isolated endurance protocols (e.g. Filipovic et al. 2019; Mathes et al. 2017) rarely results in significant differences. In contrast, a study (Amaro-Gahete et al. 2018) with recreational runners who replaced a running session with a WB-EMS session with varying impulse intensity (12–90 Hz) reported significant effects on running performance, VO_2max and running economy after 6 weeks of intervention.

> In summary, evidence for positive effects of standard WB-EMS application vs. inactive control is rather limited for endurance parameters—at least in healthy cohorts. Nevertheless, some studies reported beneficial effects of WB-EMS in older, functionally limited cohorts.

Of importance, although the effect of WB-EMS superimposed on endurance exercise is still under review, endurance athletes benefit from standard WB-EMS by preventing complaints and injuries of the musculoskeletal system[5] which would otherwise have to be addressed by much more time-consuming stabilization, resistance or physiotherapy programs.

[5] E.g. low back pain that is rather prominent in cycling and running disciplines.

5 Development of WB-EMS with Germany as an Blueprint for an Early Established Market

5.1 Historic Overview of the German WB-EMS Market

WB-EMS technology has its roots in Germany. Therefore, it is worthwhile to look at the German WB-EMS market (as the most developed WB-EMS market so far) to derive projections for the emergence of WB-EMS on international markets.

In 2000, the world's first WB-EMS training device, the "Bodytransformer", was developed and launched. In 2002, the German Sport University Cologne conducted the first studies to investigate the effectiveness of this new training technology. The first commercial successes of WB-EMS technology led to a quick but unregulated expansion of the German WB-EMS market and the emergence of many new equipment manufacturers. The lack of scientific knowledge about the side effects of incorrect WB-EMS use caused a high number of reports of cases of EMS-related harm at the time. After publication of the first case studies (Hong et al. 2016; Kastner et al. 2014; Stöllberger and Finsterer 2018; Teschler et al. 2016) and various media reports on negative health effects, the first calls for official regulations were published in 2016 (Malnick et al. 2016).

To ensure safe and effective application of WB-EMS technology, Kemmler et al. (Kemmler, Fröhlich, et al. 2016) published guidelines. The German Institute for Standardisation adopted these guidelines and published the first standard for commercial WB-EMS providers in 2018. In October of the same year, the already existing German standard for gyms ("DIN 33961 Fitness studio — requirements for studio equipment and operation") was extended by a part 5 ("Electromyostimulation training"). DIN 33961-5 regulates the safety requirements for commercial application of WB-EMS and its contraindications. Since the DIN standard was only an optional certification for gyms and commercial

W. Kemmler et al., *Whole-Body Electromyostimulation,* essentials,
https://doi.org/10.1007/978-3-031-56710-0_5

WB-EMS providers, legislators in Germany demanded a binding standard for commercial use. In 2019, the Federal Ministry for the Environment, Nature Conservation, Nuclear Safety and Consumer Protection (BMU) published the revised German Radiation Protection Statutes (BMU 2019), which under article 4 and in addition to ultrasound and laser applications, included non-medical application of electromagnetic fields. With this revision, WB-EMS ("applications of non-ionizing radiation to humans" — NiSV) was effectively incorporated into the statutes. With effect from December 31, 2022 the NiSV regulates the commercial use of WB-EMS and requires EMS instructors to undergo specific training at an officially recognized education institute (BMU 2020).

The development of the German WB-EMS market is a good example of a new and initially unregulated training technology that causes harm if used incorrectly — in some cases. Eventually, those individual cases lead to a strict regulation of the entire market by legislators and authorities.

Germany was the first country with an established WB-EMS market and the dynamics there allow a prediction for international development. On the one hand, official regulations create the legal basis for further expansion of the market. This favors those providers who adhere to high quality standards and offer professional services. The German market currently reflects this development. On the other hand, there are incidents of excessive regulation that eliminate the WB-EMS market altogether. One example for this scenario can be seen in Israel. After a single accident with incorrect WB-EMS application, the Israeli Ministry for Health installed restrictive regulations for commercial WB-EMS that make it almost impossible for training providers to offer WB-EMS (Health 2016).

5.2 Status Quo of the German Market

The annual survey of the German fitness industry provides information on the market penetration of WB-EMS (DSSV 2023). The data are broken down into single companies (facilities larger than 200 m^2), chain companies (five or more facilities larger than 200 m^2) and micro companies (200 m^2 or smaller) (Fig. 5.1).

On December 31, 2022, 15.3% of all single and chain companies were offering WB-EMS as an addition to conventional fitness training (strength, cardio and group training). The survey also shows that the number of facilities offering WB-EMS has stagnated in recent years for individual companies and decreased for chain companies. This suggests that this segment of the WB-EMS market is saturated and little change can be expected in the future. The extent to which the official regulations (NiSV) have affected the German fitness market cannot be

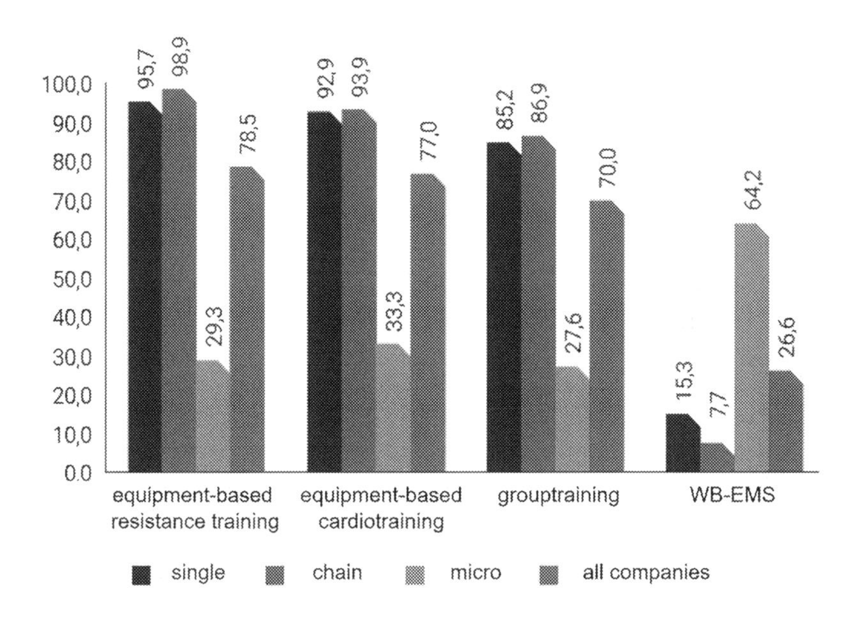

Fig. 5.1 Percentage of health clubs and gyms offering WB-EMS (DSSV, 2023)

assessed as there are no objective data on this. With 64.2% of all micro companies offering WB-EMS, the situation in this market segment is completely different from that of single and chain companies. Since the situation of micro companies was analysed in detail for the first time in 2017, a constant growth of WB-EMS providers in this segment has been observed (DSSV 2017). These figures indicate that WB-EMS for micro companies is expected to continue an upward trend.

For chain companies, the market study revealed that a mere 0.7% of the surveyed companies had invested in WB-EMS equipment in the previous year. This compares with investment in WB-EMS equipment by 59.1% of micro companies in the previous year.

5.3 The Evolution of a Business Model

The current report on the German fitness industry as well as the development of the last few years indicate that there is a trend towards a business model that focuses on WB-EMS. There are more and more micro companies offering exclusively WB-EMS. The technical term for this business model is "Special Interest Gym". In 2023, of the 9149 gyms in Germany 1434 offered only WB-EMS, which is 15.7% of all gyms (DSSV 2023). On average, the equipment in these gyms consists of two WB-EMS training devices. With an average facility size of 110 m^2, these micro companies operate on relatively small premises. Typically, clients in these gyms are thoroughly instructed and supervised by the trainer while exercising with a maximum of two clients per trainer (ibid.). According to industry reports (DSSV 2023), the overall annual turnover in the WB-EMS focused micro companies segment is 200 million Euro per year and the average turnover per company is 136,000 Euro per year. The average client generates a revenue of 98.51 Euro membership fee and additional sales (e.g. training clothes) of 14.5% per month. This is clearly above the industry average of 44.86 Euro monthly membership fee. Consumers are obviously willing to pay a higher price for personal training on WBS-EMS devices.

Interestingly, COVID-19 does not seem to have impacted this market significantly in that the expansion of this business model did not decline between 2020 and 2022. Furthermore, the NiSV regulations in 2020 did not coincide with any changes in the WB-EMS micro companies market. Advantages of the business model are the low operating costs and investment in relation to high revenues from membership fees and additional sales. Micro companies focussing on WB-EMS are established in the German fitness market and recent trends point to further growth of this market segment.

Another trend that can be observed in Germany is the growing number of companies focusing on medical WB-EMS, which is essentially training for prevention or therapy (DSSV 2023). Medical WB-EMS reaches a target group that would not participate in conventional fitness training. In particular, people with severe health or functional limitations are often unable to tolerate the mechanical loads of conventional fitness training. The training concept in micro companies often provides for close supervision by the training staff, thus qualifying them for medical WB-EMS. Clients interested in WB-EMS are often willing to invest more money for health improvements. This leads to potentially higher membership fees in this area.

The micro company business model, coupled with medical WB-EMS, is therefore a potential blueprint for markets in other countries. The following key points give a summary of this business model:

- Small facility sizes (on average: 110 m^2); low operating costs
- Focus on medical WB-EMS training
- Few WB-EMS devices per facility; low investment costs
- Focus on personal training; high predictability of human resources
- Personal training concepts justifies high membership fees

> Undoubtedly, the mandatory qualification of WB-EMS trainers in particular is an important quality criterion for the WB-EMS sector and increases customer safety. Nevertheless, the strict regulations, the cost of their implementation and the short timeframe for their realization represent a challenge for the WB-EMS market.

What you can take away from this *essential*

- WB-EMS is a time-efficient and safe training technology particularly suitable for people with low time resources or limitations that prevent conventional exercise. Apart from significant effects on fitness, function and body composition, WB-EMS demonstrated favorable effects on a large variety of health-related outcomes. In its present application with low volume and moderate to high intensity, WB EMS can be roughly considered as a resistance type exercise.

W. Kemmler et al., *Whole-Body Electromyostimulation*, essentials,
https://doi.org/10.1007/978-3-031-56710-0

References

Akçay, N., Güney, H., Kaplan, S., & Akgül, M. (2022). Electromyostimulation Exercise with Diet Program is More Effective on Body Composition than its Exercise without Diet. *MJSS, 4*, 814–822.

Aldayel, A., Jubeau, M., McGuigan, M., & Nosaka, K. (2010). Comparison between alternating and pulsed current electrical muscle stimulation for muscle and systemic acute responses. *J Appl Physiol (1985), 109* (3), 735–744.

Amaro-Gahete, F. J., De-la, O. A., Jurado-Fasoli, L., Ruiz, J. R., Castillo, M. J., & Gutierrez, A. (2019). Effects of different exercise training programs on body composition: A randomized control trial. *Scand J Med Sci Sports, 29* (7), 968–979.

Amaro-Gahete, F. J., De-la, O. A., Sanchez-Delgado, G., Robles-Gonzalez, L., Jurado-Fasoli, L., Ruiz, J. R., et al. (2018). Whole-Body Electromyostimulation Improves Performance-Related Parameters in Runners. *Front Physiol, 9*, 1576.

Beier, M., Schoene, D., Kohl, M., von Stengel, S., Uder, M., & Kemmler, W. (2024). Non-athletic cohorts addressed by longitudinal whole-body electromyostimulation trials—An evidence map. Sensors (Basel), *24*, 972.

Bellia, A., Ruscello, B., Bolognino, R., Briotti, G., Gabrielli, P. R., Silvestri, A., et al. (2020). Whole-body Electromyostimulation plus Caloric Restriction in Metabolic Syndrome. *Int J Sports Med, 41* (11), 751–758.

Berger, J., Fröhlich, M., & Kemmler, W. (2022). WB-EMS Market Development—Perspectives and Threats. *Int. J. Environ. Res. Public Health, 19*, 14211.

Berger, J., Ludwig, O., Becker, S., Backfisch, M., Kemmler, W., & Fröhlich, M. (2020). Effects of an Impulse Frequency Dependent 10-Week Whole-body Electromyostimulation Training Program on Specific Sport Performance Parameters. *J Sports Sci Med, 19* (2), 271–281.

Bircan, C., Senocak, O., Peker, O., Kaya, A., Tamci, S. A., Gulbahar, S., et al. (2002). Efficacy of two forms of electrical stimulation in increasing quadriceps strength: a randomized controlled trial. *Clin Rehabil, 16* (2), 194–199.

Bloeckl, J., Raps, S., Weineck, M., Kob, R., Bertsch, T., Kemmler, W., et al. (2022). Feasibility and Safety of Whole-Body Electromyostimulation in Frail Older People-A Pilot Trial. *Front Physiol, 13*, 856681.

W. Kemmler et al., *Whole-Body Electromyostimulation*, essentials,
https://doi.org/10.1007/978-3-031-56710-0

BMU (Ed.). (2019). *Verordnung zum Schutz vor schädlichen Wirkungen nichtionisierender Strahlung bei der Anwendung am Menschen (NiSV)* (Vol. Teil I Nr. 41). Bonn: Bundesanzeiger Verlag.

BMU (Ed.). (2020). *Anforderungen an den Erwerb der Fachkunde für Anwendungen nichtionisierender Strahlungsquellen am Menschen*. Bonn: Bundesanzeiger Verlag.

Borg, G., & Borg, E. (2010). *The Borg CR Scales® Folder.*Unpublished manuscript, Hasselby, Sweden.

Cruz-Jentoft, A. J., Baeyens, J. P., Bauer, J. M., Boirie, Y., Cederholm, T., Landi, F., et al. (2010). Sarcopenia: European consensus on definition and diagnosis: Report of the European Working Group on Sarcopenia in Older People. *Age Ageing, 39* (4), 412–423.

Cruz-Jentoft, A. J., Bahat, G., Bauer, J., Boirie, Y., Bruyere, O., Cederholm, T., et al. (2019). Sarcopenia: revised European consensus on definition and diagnosis. *Age Ageing, 48* (1), 16–31.

de la Cámara Serrano, M., Pardos, A. I., & Veiga Ó, L. (2018). Effectiveness evaluation of whole-body electromyostimulation as a postexercise recovery method. *J Sports Med Phys Fitness, 58* (12), 1800–1807.

Deutsches Institut für Normung. (2020). DIN 33402: Ergonomie - Körpermaße des Menschen. Teil 2: Werte. Berlin: Beuth-Verlag.

di Cagno, A., Buonsenso, A., Centorbi, M., Manni, L., Di Costanzo, A., Casazza, G., et al. (2023). Whole body-electromyostimulation effects on serum biomarkers, physical performances and fatigue in Parkinson's patients: A randomized controlled trial. *Front Aging Neurosci, 15*, 1086487.

DSSV. (2017). *Arbeitgeberverband deutscher Fitness- und Gesundheits-Anlagen (2017). Eckdaten der deutschen Fitness-Wirtschaft 2017*. Hamburg: DSSV.

DSSV. (2023). *Arbeitgeberverband deutscher Fitness- und Gesundheits-Anlagen (2023). Eckdaten '23 der deutschen Fitness-Wirtschaft*. Hamburg: DSSV.

Edel, H. (1991). *Fibel der Elektrodiagnostik und Elektrotherapie*. Berlin: Verlag Gesundheit.

EMS-Training.de. (2017). *EMS-Studie 2017: Die erste Endkundenbefragung*. Zirndorf.

Eriksen, E. F. (2010). Cellular mechanisms of bone remodeling. *Rev Endocr Metab Disord, 11* (4), 219–227.

Filipovic, A., Bizjak, D., Tomschi, F., Bloch, W., & Grau, M. (2019). Influence of Whole-Body Electrostimulation on the Deformability of Density-Separated Red Blood Cells in Soccer Players. *Front Physiol, 10*, 548.

Finsterer, J., & Stollberger, C. (2015). Severe rhabdomyolysis after MIHA-bodytec(R) electrostimulation with previous mild hyper-CK-emia and noncompaction. *Int J Cardiol, 180*, 100–102.

Fritzsche, D., Fruend, A., Schenk, S., Mellwig, K., Keinöder, H., Gummert, J., et al. (2010). Elektromyostimulation (EMS) bei kardiologischen Patienten. Wird das EMS-Training bedeutsam für die Sekundärprävention? *Herz, 35* (1), 34–40.

Habich, I. (2015). Muskelkraft durch EMS-Training: Gefährliche Stromstöße [Electronische Version]. *Spiegel Online*. Zugriff.

Hamada, R., Sato, S., Miyasaka, J., Murao, M., Matsushita, M., Kajimoto, T., et al. (2023). Belt Electrode-Skeletal Muscle Electrical Stimulation During Early Hematopoietic Post-Transplantation To Prevent Skeletal Muscle Atrophy and Weakness. *Transplant Cell Ther, 29* (1), 51 e51–51 e57.

Health, S. o. I. M. o. (2016). Ministry of Health has taken actions to regulate the use of EMS devices. Retrieved 31.08.2023, 2023

Hettchen, M., Glockler, K., von Stengel, S., Piechele, A., Lotzerich, H., Kohl, M., et al. (2019). Effects of Compression Tights on Recovery Parameters after Exercise Induced Muscle Damage: A Randomized Controlled Crossover Study. *Evid Based Complement Alternat Med, 2019*, 5698460.

Homma, M., Miura, M., Hirayama, Y., Takahashi, T., Miura, T., Yoshida, N., et al. (2022). Belt Electrode-Skeletal Muscle Electrical Stimulation in Older Hemodialysis Patients with Reduced Physical Activity: A Randomized Controlled Pilot Study. *J Clin Med, 11* (20), 6170.

Hong, J. Y., Oh, J. H., & Shin, J. H. (2016). Rhabdomyolysis caused by knee push-ups with whole body electromyostimulation. *Br J Hosp Med, 77* (9), 542–543.

Houdijk, A. P. J., Bos, N., Verduin, W. M., Hijdendaal, M. M., & Zwartkruis, M. A. L. (2022). Visceral fat loss by whole-body electromyostimulation is attenuated in male and absent in female older Non-Insulin-Dependent diabetes patients. *Endocrinol Diabetes Metab, 5* (6), e377.

Jee, Y.-S. (2019). The effect of high-impulse-electromyostimulation on adipokine profiles, body composition and strength: A pilot study. *J Isokinetics, 27* (3), 163–176.

Kastner, A., Braun, M., & Meyer, T. (2014). Two Cases of Rhabdomyolysis After Training With Electromyostimulation by 2 Young Male Professional Soccer Players. *Clin J Sport Med, 25* (6), 71–73.

Kemmler, W., Bebenek, M., Engelke, K., & von Stengel, S. (2014). Impact of whole-body electromyostimulation on body composition in elderly women at risk for sarcopenia: the Training and ElectroStimulation Trial (TEST-III). *Age (Dordr), 36* (1), 395–406.

Kemmler, W., Birlauf, A., & von Stengel, S. (2010). Einfluss von Ganzkörper-Elektromyostimulation auf das Metabolische Syndrom bei älteren Männern mit metabolischem Syndrom. *Dtsch Z Sportmed, 61* (5), 117–123.

Kemmler, W., Fröhlich, M., von Stengel, S., & Kleinöder, H. (2016). Whole-Body Electromyostimulation – The Need for Common Sense! Rationale and Guideline for a Safe and Effective Training. *Dtsch Z Sportmed, 67* (9), 218–221.

Kemmler, W., Fröhlich, M., Ludwig, O., Eifler, C., von Stengel, S., Willert, S., et al. (2023). Position statement and updated international guideline for safe and effective whole-body electromyostimulation training-the need for common sense in WB-EMS application. *Front Physiol, 14*, 1174103.

Kemmler, W., Grimm, A., Bebenek, M., Kohl, M., & von Stengel, S. (2018). Effects of Combined Whole-Body Electromyostimulation and Protein Supplementation on Local and Overall Muscle/Fat Distribution in Older Men with Sarcopenic Obesity: The Randomized Controlled Franconia Sarcopenic Obesity (FranSO) Study. *Calcif Tissue Int, 103* (3), 266–277.

Kemmler, W., Kleinoder, H., & Fröhlich, M. (2020). Editorial: Whole-Body Electromyostimulation: A Training Technology to Improve Health and Performance in Humans? *Front Physiol, 11*, 523.

Kemmler, W., Kohl, M., & S., v. S. (2016). Effects of High Intensity Resistance Training versus Whole-body Electromyostimulation on cardiometabolic risk factors in untrained middle aged males. A randomized controlled trial. *J Sports Res, 3* (2), 44–55.

Kemmler , W., Shojaa, M., Steele, J., Berger, J., Fröhlich, M., Schoene, D., et al. (2021). Efficacy of Whole-Body Electromyostimulation (WB-EMS) on body composition and muscle strength in non-athletic adults. A systematic review and meta-analysis. *Front Physiol, 12:640657.*

Kemmler, W., Teschler, M., Bebenek, M., & von Stengel, S. (2015). (Very) high Creatinkinase concentration after exertional whole-body electromyostimulation application: health risks and longitudinal adaptations. *Wien Med Wochenschr, 165* (21), 427–435.

Kemmler, W., Teschler, M., & Von Stengel, S. (2015). Effekt von Ganzkörper-Elektromyostimulation – „A series of studies". *Osteologie 23* (1), 20–29.

Kemmler, W., Teschler, M., Weissenfels, A., Bebenek, M., Fröhlich, M., Kohl, M., et al. (2016). Effects of Whole-Body Electromyostimulation versus High-Intensity Resistance Exercise on Body Composition and Strength: A Randomized Controlled Study. *Evid Based Complement Alternat Med, 2016*, 9236809.

Kemmler, W., Teschler, M., Weissenfels, A., Bebenek, M., von Stengel, S., Kohl, M., et al. (2016). Whole-body electromyostimulation to fight sarcopenic obesity in community-dwelling older women at risk. Results of the randomized controlled FORMOsA-sarcopenic obesity study. *Osteoporos Int, 27* (11), 3261–3270.

Kemmler, W., & von Stengel, S. (2012). Alternative Exercise Technologies to Fight against Sarcopenia at Old Age: A Series of Studies and Review. *J Aging Res, 2012*, 109013.

Kemmler, W., von Stengel, S., Kohl, M., Rohleder, N., Bertsch, T., Sieber, C. C., et al. (2020). Safety of a Combined WB-EMS and High-Protein Diet Intervention in Sarcopenic Obese Elderly Men. *Clin Interv Aging, 15*, 953–967.

Kemmler, W., Weissenfels, A., Bebenek, M., Fröhlich, M., Kleinoeder, H., Kohl, M., et al. (2017). Effects of Whole-Body-Electromyostimulation (WB-EMS) on low back pain in people with chronic unspecific dorsal pain – a meta-analysis of individual patient data from randomized controlled WB-EMS trials. *Evid Based Complement Alternat Med, Article ID 8480429*, doi: https://doi.org/10.1155/2017/8480429.

Kemmler, W., Weissenfels, A., Teschler, M., Willert, S., Bebenek, M., Shojaa, M., et al. (2017). Whole-body Electromyostimulation and protein supplementation favorably affect Sarcopenic Obesity in community-dwelling older men at risk. The Randomized Controlled FranSO Study. *Clin Interv Aging, 12*, 1503–1513.

Kemmler, W., Kohl, M., von Stengel, S., Willert, S., Kast, S., Uder, M. (2024) Effects of whole-body electromyostimulation with different impulse intensity on blood pressure changes in hyper- and normotensive overweight people. A pilot study. *Front Physiol. 15*, 1349750.

Kemmler, W., Weissenfels, A., Willert, S., Fröhlich, M., Ludwig, O., Berger, J., et al. (2019). Recommended Contraindications for the Use of Non-Medical WB-Electromyostimulation. *Dtsch Z Sportmed, 70* (11), 278–281.

Kim, J., & Jee, Y. (2020a). Aerobic Dance Wore with EMS Suit Improves Fatness and Biomarkers of Obese Elderly Women, *Preprints*: Preprints.

Kim, J., & Jee, Y. (2020b). EMS-effect of Exercises with Music on Fatness and Biomarkers of Obese Elderly Women. *Medicina (Kaunas), 56* (4), 156.

Kim, K., Eun, D., & Jee, Y. S. (2021). Higher Impulse Electromyostimulation Contributes to Psychological Satisfaction and Physical Development in Healthy Men. *Medicina (Kaunas), 57* (3).

Koch, A. J., Pereira, R., & Machado, M. (2014). The creatine kinase response to resistance exercise. *J Musculoskelet Neuronal Interact, 14* (1), 68–77.

Konrad, K. L., Baeyens, J.-P., Birkenmaier, C., Ranker, A. H., Widmann, J., Leukert, J., et al. (2020). The effects of whole-body electromyostimulation (WB-EMS) in comparison to a multimodal treatment concept in patients with non-specific chronic back pain—A prospective clinical intervention study. *PloS one, 15* (8), e0236780.

Konrad, K. L., Weissenfels, A., Baeyens, J., Kemmler, W., & Wegener, B. (2023). Whole-body electromyostimulation (WB-EMS) on non-specific chronic back pain: Patients with what degree of pain intensity benefit the most? *Front Physiol, submitted.*

Laufer, Y., & Elboim, M. (2008). Effect of burst frequency and duration of kilohertz-frequency alternating currents and of low-frequency pulsed currents on strength of contraction, muscle fatigue, and perceived discomfort. *Phys Ther, 88* (10), 1167–1176.

Le, Y. E., Schoene, D., Kohl, M., von Stengel, S., Uder, M., & Kemmler, W. (2024). Outcomes addressed by longitudinal whole-body electromyostimulation trials in middle aged-older adults – An evidence map. *PLOS ONE, submitted.*

Ludwig, O., Berger, J., Schuh, T., Backfisch, M., Becker, S., & Fröhlich, M. (2020). Can A Superimposed Whole-Body Electromyostimulation Intervention Enhance the Effects of a 10-Week Athletic Strength Training in Youth Elite Soccer Players? *J Sports Sci Med, 19* (3), 535–546.

Lukashevich, U. A., Ponomarev, V.V., Tarasevich, M.I., Zhivolupov, S.A. (2020). Functional reciprocal neuromuscular electric stimulation in adaptive kinesitherapy in post-stress patients. *Science & Healthcare, 22* (3), 80–88.

Malnick, S. D., Band, Y., Alin, P., & Maffiuletti, N. A. (2016). It's time to regulate the use of whole body electrical stimulation. *BMJ, 352*, i1693.

Mathes, S., Lehnen, N., Link, T., Bloch, W., Mester, J., & Wahl, P. (2017). Chronic effects of superimposed electromyostimulation during cycling on aerobic and anaerobic capacity. *Eur J Appl Physiol, 117* (5), 881–892.

Micke, F., Weissenfels, A., Wirtz, N., Von Stengel, S., Dörmann, U., Kohl, M., et al. (2021). Similar Pain Intensity Reductions and Trunk Strength Improvements following Whole-Body Electromyostimulation vs. Whole-Body Vibration vs. Conventional Back-Strengthening Training in Chronic Non-specific Low Back Pain Patients: A 3-armed randomized controlled trial. *Front Physiol, 13* (12), 664991.

Müllerová, M., Vaculíková, P., Potúčková, A., Struhár, I., & Balousová, D. N. (2022). Impact of Whole-Body Electromyostimulation and Resistance Training Programme on Strength Parameters and Body Composition in Group of Elderly Women at Risk of Sarcopenia. *Studia sportiva, 16* (2), 292–304.

Noguchi, Y., Hirano, H., Mizutani, C., Ito, T., & Kawamura, N. (2018). [The effect of electrical stimulation of the skeletal muscles with belt electrodes during hemodialysis on the physical function of hemodialysis patients]. *Journal of Dialysis Society, 51* (1), 87–91.

Paillard, T. (2021). Sensory electrical stimulation and postural balance: a comprehensive review. *Eur J Appl Physiol, 121* (12), 3261–3281.

Pano-Rodriguez, A., Beltran-Garrido, J. V., Hernandez-Gonzalez, V., & Reverter-Masia, J. (2020). Effects of Whole-Body Electromyostimulation on Physical Fitness in Post-menopausal Women: A Randomized Controlled Trial. *Sensors (Basel), 20* (5).

Park, S., Min, S., Park, S. H., Yoo, J., & Jee, Y. S. (2021). Influence of Isometric Exercise Combined With Electromyostimulation on Inflammatory Cytokine Levels, Muscle

Strength, and Knee Joint Function in Elderly Women With Early Knee Osteoarthritis. *Front Physiol, 12*, 688260.

Park, S., Park, J., Ham, J. A., & Jee, Y. (2021). Effects of aerobic dance with electrical stimulant on body composition and radiological circumference of obese elderly women. *Gazzetta Medica Italiana Archivio per le Scienze Mediche, 180* (3), 87–95.

Park, W., Lee, J., Hong, K., Park, H. Y., Park, S., Kim, N., et al. (2023). Protein-Added Healthy Lunch-Boxes Combined with Exercise for Improving Physical Fitness and Vascular Function in Pre-Frail Older Women: A Community-Based Randomized Controlled Trial. *Clin Interv Aging, 18*, 13–27.

Pinar, S., Kaya, F., Bicer, B., Erzeybek, M. S., & Cotuk, H. B. (2012). Different recovery methods and muscle performance after exhausting exercise: comparison of the effects of electrical muscle stimulation and massage. *Biol Sport, 29* (4), 269–275.

Pinfildi, C. E., Andraus, R. A. C., Iida, L. M., & Prado, R. P. (2018). Neuromuscular Electrical Stimulation of Medium and Low Frequency on the Quadriceps Femoris. *Acta Ortop Bras, 26* (5), 346–349.

Querol, F., Gallach, J. E., Toca-Herrera, J. L., Gomis, M., & Gonzalez, L. M. (2006). Surface electrical stimulation of the quadriceps femoris in patients affected by haemophilia A. *Haemophilia, 12* (6), 629–632.

Reljic, D., Dieterich, W., Herrmann, H. J., Neurath, M. F., & Zopf, Y. (2022). "HIIT the Inflammation": Comparative Effects of Low-Volume Interval Training and Resistance Exercises on Inflammatory Indices in Obese Metabolic Syndrome Patients Undergoing Caloric Restriction. *Nutrients, 14* (10), 1996.

Reljic, D., Herrmann, H. J., Neurath, M. F., & Zopf, Y. (2021). Iron Beats Electricity: Resistance Training but Not Whole-Body Electromyostimulation Improves Cardiometabolic Health in Obese Metabolic Syndrome Patients during Caloric Restriction-A Randomized-Controlled Study. *Nutrients, 13* (5), 1640.

Reljic, D., Konturek, P. C., Herrmann, H. J., Neurath, M. F., & Zopf, Y. (2020). Effects of whole-body electromyostimulation exercise and caloric restriction on cardiometabolic risk profile and muscle strength in obese women with the metabolic syndrome: a pilot study. *J Physiol Pharmacol, 71* (1), 89–98.

Ricci, P. A., Di Thommazo-Luporini, L., Jurgensen, S. P., Andre, L. D., Haddad, G. F., Arena, R., et al. (2020). Effects of Whole-Body Electromyostimulation Associated with Dynamic Exercise on Functional Capacity and Heart Rate Variability After Bariatric Surgery: a Randomized, Double-Blind, and Sham-Controlled Trial. *Obes Surg, 30*, 3862–3871.

Roos, E. M., & Lohmander, L. S. (2003). The Knee injury and Osteoarthritis Outcome Score (KOOS): from joint injury to osteoarthritis. *Health Qual Life Outcomes, 1*, 64.

Sánchez-Infante, J., Bravo-Sáncheza, A., Abiánb, P., Estebana, P., Jimeneza, J., & Abián-Vicén, J. (2020). The influence of whole-body electromyostimulation training in middle-aged women. *Isokinet Exerc Sci 1*, 1–9.

Schink, K., Gassner, H., Reljic, D., Herrmann, H. J., Kemmler, W., Schwappacher, R., et al. (2020). Assessment of gait parameters and physical function in patients with advanced cancer participating in a 12-week exercise and nutrition programme: A controlled clinical trial. *Eur J Cancer Care (Engl), 29* (2), e13199.

Schink, K., Herrmann, H. J., Schwappacher, R., Meyer, J., Orlemann, T., Waldmann, E., et al. (2018). Effects of whole-body electromyostimulation combined with individualized nutritional support on body composition in patients with advanced cancer: a controlled pilot trial. *BMC Cancer, 18* (1), 886.

Schwappacher, R., Dieterich, W., Reljic, D., Pilarsky, C., Mukhopadhyay, D., Chang, D. K., et al. (2021). Muscle-Derived Cytokines Reduce Growth, Viability and Migratory Activity of Pancreatic Cancer Cells. *Cancers (Basel), 13* (15), 3820.

Schwappacher, R., Schink, K., Sologub, S., Dieterich, W., Reljic, D., Friedrich, O., et al. (2020). Physical activity and advanced cancer: evidence of exercise-sensitive genes regulating prostate cancer cell proliferation and apoptosis. *J Physiol, 598* (18), 3871–3889.

Silvestri, A., Ruscello, B., Rosazza, C., Briotti, G., Gabrielli, P. R., Tudisco, C., et al. (2023). Acute Effects of Whole-Body Electrostimulation Combined with Stretching on Lower Back Pain. *Int J Sports Med, 44* (11), 820–829.

Stefanovska, A., & Vodovnik, L. (1985). Change in muscle force following electrical stimulation. Dependence on stimulation waveform and frequency. *Scand J Rehabil Med, 17* (3), 141–146.

Stephan, H., Wehmeier, U. F., Forster, T., Tomschi, F., & Hilberg, T. (2023). Additional Active Movements Are Not Required for Strength Gains in the Untrained during Short-Term Whole-Body Electromyostimulation Training. *Healthcare (Basel), 11* (5), 741.

Stöllberger, C., & Finsterer, J. (2018). Acute myopathy as a side effect of electromyostimulation. Letter to the editor. *WMW, 169*, 181–182.

Strahlenschutzkommission (Ed.). (2019). *Anwendung elektrischer, magnetischer und elektromagnetischer Felder (EMF) zu nichtmedizinischen Zwecken am Menschen. Empfehlung der Strahlenschutzkommission mit wissenschaftlicher Begründung.* Bonn.

Sulprizio, M., & Kleinert, J. (2016). *Sport in der Schwangerschaft. Leitfaden für die geburtshilfliche und gynäkologische Beratung [Exercise during pregnancy].* Berlin: Springer-Verlag.

Teschler, M., Heimer, M., Schmitz, B., Kemmler, W., & Mooren, F. C. (2021). Four weeks of electromyostimulation improves muscle function and strength in sarcopenic patients: a three-arm parallel randomized trial. *J Cachexia Sarcopenia Muscle, 12* (4), 843–854.

Teschler, M., Wassermann, A., Weissenfels, A., Fröhlich, M., Kohl, M., Bebenek, M., et al. (2018). Short time effect of a single session of intense whole-body electromyostimulation on energy expenditure. A contribution to fat reduction? *Appl Physiol Nutr Metab, 43* (5), 528–530.

Teschler, M., Weissenfels, A., Bebenek, M., Fröhlich, M., Kohl, M., von Stengel, S., et al. (2016). Very high creatine kinase CK levels after WB_EMS. Are there implications for health. *Int J Clin Exp Med 9*(11), 22841–22850.

Tsurumi, T., Tamura, Y., Nakatani, Y., Furuya, T., Tamiya, H., Terashima, M., et al. (2022). Neuromuscular Electrical Stimulation during Hemodialysis Suppresses Postprandial Hyperglycemia in Patients with End-Stage Diabetic Kidney Disease: A Crossover Controlled Trial. *J Clin Med, 11* (21), 6239.

Vaculikova, P. P., A. Kotkova, M. Struhar, I. Balousova, D. N. (2022). Impact of Whole-Body Electromyostimulation and Resistance Training on Bone Mineral Density in women at risk for Osteopororosis. *IJPESS*, 69–79.

Vaculikova, P. P., A. Kotkova, M. Struhar, I. Balousova, D. Rozsypal, R. (2023). Impact of Whole-Body Electromyostimulation and Resistance Training on the Level of Functional Fitness in Elderly Women. *Studia Sportiva, 16* (2), 115–126.

van Buuren, F., Horstkotte, D., Mellwig, K., Fruend, A., Bogunovic, N., Dimitriadis, Z., et al. (2015). Electrical Myostimulation (EMS) Improves Glucose Metabolism and Oxygen Uptake in Type 2 Diabetes Mellitus Patients — Results from the EMS Study. *Diabetes Technol Ther, 17* (6), 413–419.

van Buuren, F., Mellwig, K. P., Prinz, C., Korber, B., Frund, A., Fritzsche, D., et al. (2013). Electrical myostimulation improves left ventricular function and peak oxygen consumption in patients with chronic heart failure: results from the exEMS study comparing different stimulation strategies. *Clin Res Cardiol, 102* (7), 523–534.

van Kerkhof, P. (2022). *Evidenzbasierte Elektrotherapie [evidence-based electrotherapy]*. Berlin: Springer-Verlag.

Visser, M., Goodpaster, B. H., Kritchevsky, S. B., Newman, A. B., Nevitt, M., Rubin, S. M., et al. (2005). Muscle mass, muscle strength, and muscle fat infiltration as predictors of incident mobility limitations in well-functioning older persons. *J Gerontol A Biol Sci Med Sci, 60* (3), 324–333.

von Stengel, S., Bebenek, M., Engelke, K., & Kemmler, W. (2015). Whole-Body Electromyostimulation to Fight Osteopenia in Elderly Females: The Randomized Controlled Training and Electrostimulation Trial (TEST-III). *J Osteoporos, 2015*, 643520.

Weissenfels, A., Teschler, M., Willert, S., Hettchen, M., Fröhlich, M., Kleinoder, H., et al. (2018). Effects of whole-body electromyostimulation on chronic nonspecific low back pain in adults: a randomized controlled study. *J Pain Res, 11*, 1949–1957.

Willert, S., Weissenfels, A., Kohl, M., von Stengel, S., Fröhlich, M., Kleinöder, H., et al. (2019). Effects of Whole-Body Electromyostimulation (WB-EMS) on the energy-restriction-induced reduction of muscle mass during intended weight loss. *Front Physiol* (10), 1012.

Wirtz, N., Dormann, U., Micke, F., Filipovic, A., Kleinoder, H., & Donath, L. (2019). Effects of Whole-Body Electromyostimulation on Strength-, Sprint-, and Jump Performance in Moderately Trained Young Adults: A Mini-Meta-Analysis of Five Homogenous RCTs of Our Work Group. *Front Physiol, 10*, 1336.

Wittmann, K., Sieber, C., von Stengel, S., Kohl, M., Freiberger, E., Jakob, F., et al. (2016). Impact of whole body electromyostimulation on cardiometabolic risk factors in older women with sarcopenic obesity: the randomized controlled FORMOsA-sarcopenic obesity study. *Clin Interv Aging, 11*, 1697–1706.

Zitzmann, A. L., Shojaa, M., Kast, S., Kohl, M., von Stengel, S., Borucki, D., et al. (2022). The effect of different training frequency on bone mineral density in older adults. A comparative systematic review and meta-analysis. *Bone, 154*, 116230.

Printed in the United States
by Baker & Taylor Publisher Services